A GUIDE
AUSTRA_____
MOTHS

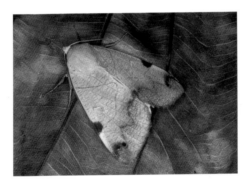

PAUL ZBOROWSKI AND TED EDWARDS

CSIRO
PUBLISHING

National Library of Australia Cataloguing-in-Publication entry
Zborowski, Paul.
    A guide to Australian moths.

    Includes index.
    ISBN 9 78064309 1597.

    1. Moths – Australia – Identification. I. Edwards, Ted.
    II. Title.

    595.780994

Published by
CSIRO Publishing
150 Oxford Street (PO Box 1139)
Collingwood VIC 3066
Australia

Telephone:	+61 3 9662 7666
Local call:	1300 788 000 (Australia only)
Fax:	+61 3 9662 7555
Email:	publishing.sales@csiro.au
Web site:	www.publish.csiro.au

Front cover
Xenogenes gloriosa. Photo by Paul Zborowski.
Back cover
From left to right: *Anthela ocellata*, larva of *Aeolochroma metarhodata*, *Ophiusa tirhaca*. All photos by Paul Zborowski.

Set in 9.5/12 Minion
Cover and text design by James Kelly
Typeset by James Kelly
Printed in China by Bookbuilders

FOREWORD

Australia has a wealth of moth species, many of which remain to be identified. While most people are familiar with the closely related butterflies which fly during the day, are frequently brilliantly coloured and are attracted to garden flowers, our moths are not so well understood and appreciated. There are 140 families of moths and just five families of butterflies.

Moths comprise a very significant component of the environment. From a negative point of view they include many pests of horticultural and broad-acre crops and clothes moths are familiar household pests. On a positive note moths have been very important in the biological control of weeds. The extremely widespread and threatening occurrence of Prickly Pear in southern Queensland and northern New South Wales was controlled by the introduction of a moth from Argentina, *Cactoblastis*.

Moths play a vital role in pollination and are a source of food for birds and bats. A key role of moths in forest and woodland is in the breakdown of leaf litter. Moth larvae are particularly important in digesting the tough residues derived from eucalypts and other Australian flora, in reducing the fuel load and thus the damage from bushfires, and in facilitating nutrient recycling.

A Guide to Australian Moths provides the reader with an easy way to discover what family a particular moth belongs to. It answers some of the most commonly asked questions about moths, such as 'How do moths see?' 'Why are moths attracted to light?' and 'Have any species become extinct?'

This book does not set out to be a scientific tome or to provide specific identification of the vast number of Australian moths. It is very accessible and easy to read and is illustrated with more than 400 high quality photographs, which makes identification of moth families that much easier.

A Guide to Australian Moths makes a much needed and timely contribution to the literature on Australian natural history. In displaying the huge diversity of Australian moths, the beautiful and intricate patterning of their wings and their extraordinary adaptations, it can only serve to inspire readers with respect for the natural environment and interest in their future protection.

John Landy AC MBE

CONTENTS

This Clear-winged Hawk moth, *Cephonodes kingii*, is feeding at the flowers of *Buddleja davidii* (Buddlejaceae). It is day-flying and found in tropical and subtropical Australia and on rare occasions is seen as far south as Canberra.

Photo: Robert Luttrell

PREFACE

Australia has somewhere between 20 000 to 30 000 species of moths (a number comparable to the number of flowering plants) but only about 400 butterflies. Our butterflies therefore make up less than 2% of the insect order Lepidoptera, the moths and butterflies. Unlike North America, which has only 11 000 to 12 000 species of moths but over 700 butterflies, we have a very rich moth fauna and a rather depauperate butterfly fauna as the butterflies have not adapted well to the arid conditions in Australia.

This book contains about 400 images so can include no more than about 2% of the Australian moths. So the reader is likely to find in the book few of the moths she or he sees at the porch light. We have therefore included simple lists of characteristic features to help narrow the possibilities down to one or two families, which will make it much easier to access general information from the wealth of knowledge existing about their lives, biology, interactions with other animals and plants, and with humans.

Visitors to CSIRO Entomology are fascinated by the various stories about the moths exhibited and this book attempts to bring some of these stories to a wider audience. The introductory section of the book answers the most frequently asked questions at open days. These answers are comprehensive and include related topics to give a more balanced introduction to the moths.

Although this book is not intended to be a scientific work, it does provide information about moth evolution and why they have been successful. It is about living moths rather than pinned moths and never before have so many photographs of such a diverse range of live moths been published. These photographs should allow recognition of the different families of moths without the equipment such as microscopes normally needed to see their characteristic features. In addition, a series of 13 'boxes' provides detailed information about a few moths of particular interest to many Australians. However, the book does not cover collecting, collections, techniques for rearing immature stages or give any historical perspective.

All the photographs in the book were taken by Paul Zborowski unless otherwise attributed.

Readers interested in more detailed information should consult the book, *Moths of Australia* by the late I. F. B. Common published in 1990 by Melbourne University Press. This can be obtained easily second-hand on the web. Images of pinned and set specimens of several thousand identified Australian moths may be viewed on www.ento.csiro.au/anic/moths.html. This is a wonderful identification aid but far from complete at this stage. A list in book form of all the recorded Australian moths was published by CSIRO Publishing in 1996 and entitled *A Checklist of the Moths of Australia* by E. S. Nielsen, E. D. Edwards and T. V. Rangsi. This book is also available second-hand on the web. A detailed catalogue of a few families may be found on the Australian Biological Resources Study website: www.deh.gov.au/biodiversity/abrs/online-resources/fauna/afd/group.html#lepi-doptera. Don Herbison Evans runs a site illustrating larvae at www.usyd.edu.au/macleay/larvae. David Britton also posts some moth information on the Australian Museum, Sydney website www.amonline.net.au/factsheets/#insects.

ACKNOWLEDGMENTS

We are very lucky to have had a great deal of very generous help and support from many people without whom this book could not have been produced.

We would like to thank Mrs Natalie Barnett, Dr Michael Braby, the late Dr Ian Common, Mr Fabian Douglas, Dr George Gibbs, Mr John Green, the late Mr Bob Jessop, Dr Lauri Kaila, Dr Axel Kallies, Mr Ray McInnes, Mr Peter Marriott, Dr David Rentz, Dr Don Sands, Dr Steve Shattuck, Mr John Stockard, Mr Jan Taylor, Mr Murray Upton, Dr Andreas Zwick and CSIRO Entomology for permission to use photographs they have taken or have in their care. There are a few photographs from the CSIRO Entomology photo library that we could not attribute to a photographer and we apologise to anyone whose photograph has been used with incorrect acknowledgement.

Ms Jo Cardale, Dr Max Day, Dr Marianne Horak, Dr Niels Kristensen, Dr Tosio Kumata, Mr David Lane, Dr Alice Wells and Dr Max Moulds have provided very valuable advice on the identity of some of the insects illustrated or on other problems; their help is gratefully acknowledged. Over many years colleagues at CSIRO Entomology have given unstinting support, time, advice and encouragement and especially the late Dr Ian Common, Dr Marianne Horak, Ms Vanna Rangsi and Mr Murray Upton. Help with fieldwork to locate odd families was provided by Mr Glenn Cocking, Dr Peter McQuillan,

Dr David Rentz and Dr Andreas Zwick and friends in Sydney, Mr Bart Hacobian and Mr Len Willan, have been a constant source of inspiration and enthusiasm.

Parts of the manuscript were read by Mr Glenn Cocking, Mr Lyn Craven, Dr Doug Hilton, Dr Marianne Horak and Dr Andreas Zwick, and many helpful comments were received. Dr Joanne Daly, Chief of CSIRO Entomology, has indulged one of the authors with the opportunity to write the book while holding a Post Retirement Fellowship with CSIRO Entomology and permitted the use of illustrations that appeared in *Insects of Australia*, published by Melbourne University Press.

The CSIRO Entomology photo library has been fully available and permission to reproduce many photographs is gratefully acknowledged. A great many other colleagues too numerous to mention have helped in building, maintaining and supporting the Australian National Insect Collection at CSIRO Entomology and provided much personal help and friendship over many years.

We are grateful for the early stages of the Hercules moth, which were provided by Ms Silke Weyland and the assistance provided by Ms Anja Bakker and Ms Robyn Cruse of the Australian Butterfly Sanctuary, Kuranda. Ms Ann Crabb provided enthusiastic support for the project in CSIRO Publishing and this was later taken up by Mr Nick Alexander who has provided cheerful and continuing assistance.

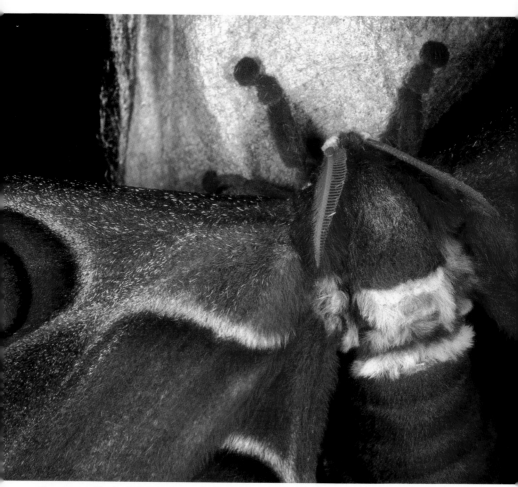

The wings and body of this Hercules moth, *Coscinocera hercules* (Saturniidae), are covered with loose scales characteristic of all moths.

INTRODUCTION

What is a moth?

Moths are insects with four wings and a long, coiled proboscis with which they suck nectar. If an insect has a coiled sucking proboscis then it must be a moth or butterfly. However, the most primitive moths, the Micropterigidae and Agathiphagidae, have chewing mouthparts inherited from their ancestors before the coiled proboscis was evolved. Many other moths have secondarily lost the proboscis (their ancestors had one but lost it) and have no functional mouthparts at all.

Moths and butterflies all have small, broad, flattened scales (like dust) on their wings and broad or hair-like scales on their bodies. These scales are often brightly coloured. A few have lost most of these scales from the wings but there are always some present. These features charac-terise the great insect order Lepidoptera (meaning 'scaly wings' in classical Greek)—the moths and butterflies. The caddisflies (Trichoptera, meaning 'hairy wings') are the order of insects most closely related to moths and some of these have small hairs on the wings but never broad, flattened scales.

What is the difference between a moth and a butterfly?

Both moths and butterflies have a coiled proboscis and scaly wings. On the family tree of Lepidoptera about 140 branches are moths and five are butterflies. The five butterfly families arise from a single branch within the moths, meaning they had a common ancestor that did not also lead to another moth group. However, even this is contested as one Central American moth family (Hedylidae) has so many butterfly-like features it may have arisen from within the butterfly line. So the difference between butterflies and moths is not great, and is comparable with the differences between moth families.

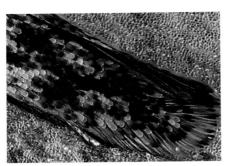

The broad, flattened wing scales of the scribbly gum moth, *Ogmograptis* sp. (Bucculatricidae).
Photo: Natalie Barnett

Most moths have antennae that are thread-like or feathery, such as this Hercules moth.

1

A butterfly, *Junonia villida,* in the family Nymphalidae, with the clubbed antennae characteristic of butterflies but rarely present in moths. Photo: Bob Jessop

Moths usually have antennae that are thread-like or feathery (pectinate). Butterflies always have antennae that have a marked club at the tip. However, there are some mostly day-flying moths that have clubbed antennae, in particular the Castniidae, and it may be that there is some broad correlation between a day-flying habit and clubbed antennae.

Most moth families have a hook-and-bristle system that helps to keep the forewings and hindwings functioning together in flight. This is done by means of a bristle and hook, called a frenulum and retinaculum (see p. 34), which is not present in butterflies where the wings work together because of a generous overlap between the forewings and hindwings. If a moth or butterfly has clubbed antennae but no hook-and-bristle system, then it must be a butterfly. Everything else is a moth. There is one nasty exception to this in the world and that is the male Australian skipper butterfly, *Euschemon rafflesia*, which has clubbed antennae and a hook and bristle.

It is very broadly true that butterflies fly during the day and moths fly at night but there are so many exceptions that this is of no practical use in distinguishing moths from butterflies. Some moths such as the Castniidae are so committed to day flight that they never move at night, and one Australian butterfly is more likely to be active at night than in the day. Many moths found in the mountains where it is very cold at night will fly in the day, and many fly in rainforests where conditions are moist. A few moth species fly for a short period in the early rays of the sun at dawn, particularly in arid Australia.

What features have made moths so successful?

All animals that do not have an internal skeleton of bones are called invertebrates, and most of these have a hard external skeleton. Many things including sea-anemones, worms, slugs and beetles are invertebrates. Some of these have jointed legs and are called arthropods. The arthropods include animals such as lobsters, prawns, spiders, mites, centipedes and insects.

Moths belong to the huge class of animals called insects. Insects are characterised by having three pairs of legs and a body in three distinct sections: a head, thorax and abdomen. Each pair of legs arises from each of the three segments of the thorax. Many insects are familiar to us because we so frequently see them in daily life: moths, flies, beetles, wasps, bugs, grasshoppers, dragonflies and silverfish are all insects. Silverfish do not have wings but all other common insects have wings unless they have been secondarily lost.

A hard external skeleton can make it very difficult to grow. Insects solve this problem by shedding their external skeleton and growing a new, larger one periodically throughout their growing stage. This method of accommodating growth means that great differences are possible in the appearance of succeeding developmental stages and the higher insects have developed a complete metamorphosis where, following the egg stage, there is a distinct larval stage then a pupal stage and finally the adult.

Moths, flies, wasps, beetles, lacewings and caddisflies and a few other less

Larvae are specialised to act as the main feeding stage in the life cycle. This is *Spodoptera picta* (Noctuidae), which feeds on lilies. Many noctuid larvae such as cutworms and budworms, which rest in sheltered places, have this general shape. The moth is illustrated on page 191.

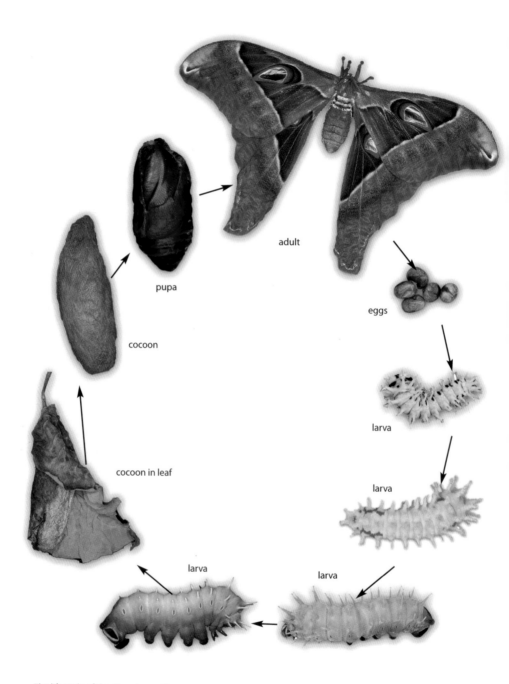

The life cycle of the Hercules moth.

This is part of a clutch of eggs of *Spilosoma glatignyi* (Arctiidae). Some moths lay eggs in batches while others lay them singly. The larvae of this species feed on a wide range of herbs and disperse after hatching. Photo: Bob Jessop

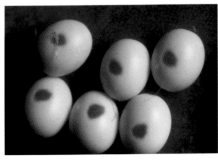

The large porcelain-like eggs of *Pararguda nasuta* (Lasiocampidae) are often laid in batches. The eggs of most moths hatch after about a fortnight but a few remain as eggs for much longer and some require rain before they will hatch. This moth belongs in the bombycoid group where the females do not feed as adults, mate very quickly and lay large eggs. They will lay eggs in captivity even without their necessary foodplant. Photo: Bob Jessop

frequently seen groups all have a complete metamorphosis. This has permitted the larval stage to become a specialised feeding, growing and hiding stage, and the adult stage to become a highly mobile flying stage specialised in making contact with the opposite sex and dispersing the eggs. The egg stage, as in all other animals, allows the adult to produce many offspring and the pupal stage is a specialised resting stage permitting the vast reorganisation

needed to change from a sedentary larva to a winged adult.

All the growth in moths takes place in the larval (caterpillar) stage, and the adult does not grow. However, many adult moths feed on nectar, fermenting fruit or sap flows that are of high-energy content and help meet the great energy demands of flight.

The complete metamorphosis allows moths to have a caterpillar stage that is relatively immobile, reclusive and able to grow on a diet that is plentiful but of low nutritive value, usually of green leaves, and at the same time to have a very mobile adult stage that is able to use foods of very high-energy value needed for flight but which are very scarce. In practice, many moths rarely feed as adults but they do drink dew. In Australia, those most seen at flowers are the Sphingidae, Noctuidae and a few families of the smaller moths. The energy needs of most are met by the stores of energy accumulated from the larval stage and by having a short but not-so-sweet adult life.

The jointed legs of arthropods make walking possible and all animals without legs have to slide, wriggle or swim. On dry land jointed legs allow much greater mobility and this is one reason why the invertebrate fauna of the dry land is so dominated by arthropods. Caterpillars seem to have lots of legs but only the legs of the first three body segments, corresponding to the thorax of the adult, are jointed. There are, in many moth larvae, another five pairs of legs called prolegs that are not jointed and are not equivalent to the basic insect jointed legs seen in adult moths. These prolegs are fleshy projections from the body and are usually tipped with a row of tough hooks with which the larva can grip the substrate securely. The development of prolegs has facilitated leaf-feeding in Lepidoptera where a larva can grip a stem with safety even in strong winds and particularly where it has laid a silken trail or prepared a silken resting pad. They are found only in moth and butterfly larvae and provide a means of distinguishing the larvae of moths from those of the main leaf-feeding rivals of moths, chrysomelid beetles and sawflies, which do not have prolegs. The prolegs are sometimes lost in some leaf-mining larvae.

A leaf-mining larva lives and feeds between the upper and lower surfaces of a leaf. The damage it causes by feeding shows as either a blotch or a blister on the leaf, or in a long serpentine 'mine', sometimes ending in a blotch.

Access to the vast low-nutritive-value food source of leaves is an enormous advantage if the larva can be inconspicuous enough. Another advantage of living in the leaves is the protection from ants and other walking predators. To find a larva an ant has to make a series of choices of which branch to walk along all the way up the tree to find the larva, and at the end the ant's chances of finding it are not very good. It is rather like the ant winning a lottery. Some adult butterflies and day-flying moths rest for the night on the ends of twigs or the tips of long grass blades for the same reason.

Another feature of insect (and moth) life is their breathing method. Most segments of the body have a spiracle, an opening that permits the entry of air into a many-branched tracheal system that penetrates finely through the body allowing oxygen access to the tissues.

This system is partly passive where the oxygen moves very slowly by diffusion but is made more efficient by the action of the flight muscles moving air within the tracheae and by the outside airflow over the spiracles. Contrast this to the familiar land vertebrate system where lungs are worked actively to obtain oxygen, which is then distributed in the circulatory system. The active vertebrate system allows the animal to maintain a high body temperature, even at rest. Moths can operate over a wide range of body temperatures but many of the larger moths such as the Hepialidae, Sphingidae and Noctuidae operate most efficiently at quite high thoracic temperatures. Some of these moths will 'shiver' to

The prolegs (false legs on the abdomen) of this larva of *Syntherata janetta* (Saturniidae) grip the stem tightly. Here the yellow hairs are probably sensory but the gripping hooks are brown and can be seen in contact with the stem. These larvae can really hang on in a wind.

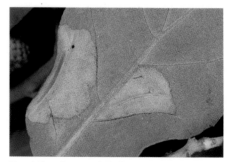

The old larval mines of *Phyllonorycter messaniella* (Gracillariidae) show in the leaves of a cork oak. Many small moths are miners in leaves as larvae, and some larger ones start life as a miner.

The leaf mine of the gracillariid *Acrocercops plebeia* in Queensland wattle, *Acacia podalyriifolia*. Photo: Bob Jessop

This hawk moth larva, in the family Sphingidae, has very prominent spiracles or breathing holes on most segments of the body.

warm the flight muscles before flying, and once active the flight muscles generate more than enough heat to warm the thorax.

Moths, however, have greater cooling options than vertebrates; most vertebrates have to literally sweat it out if they get too hot. Because oxygen is not circulated in the blood, moths can close off the blood circulation to the abdomen for a long time without ill effect. They can thus isolate the thorax so that warm-up is quicker and then, when at optimum temperature, they can progressively open vascular access to the abdomen using the abdomen as a heat sink and so maintain a stable thoracic muscle temperature usually around 35°C.

Maintaining a high thoracic temperature in flight allows a moth to have smaller, more controllable wings and to have a much higher wing-beat frequency, and so to be much more agile in the air. This allows moths to exploit the night where there are no external sources of heat and when basking (as used by reptiles and some day-flying moths and butterflies) is not possible. The vast majority of moths are very small, do not maintain a significantly high thoracic

muscle temperature, and function on what is called a very low wing loading, which allows a rather slow and slowly flapping flight. This makes the moths very vulnerable to predators but the low light levels at night make it very difficult for most predators to see well enough to prey on even slowly flying moths.

It is no accident that the main night aerial predators of moths are vertebrates with a high body temperature (including birds such as nightjars and frogmouths), or bats that have switched to sonar for prey location. Invertebrate aerial predators, such as robberflies and dragonflies, do not seem to be so active at night. As we will see later, bats have had a significant evolutionary impact on moths. Another reason why poor aerialists may fly mostly at night is that air currents can be more stable at night without the more violent convective wind currents generated by sunshine.

Moths, in contrast to other insects, have an extensive body and wing covering of scales. These may be very broad (as are those on the wings) or they may be long and hair-like (as on the thorax). These hair-like scales can provide very good insulation that helps maintain a high thoracic muscle temperature while in flight. The scales of moths also help them to escape from spiders; when encountering a sticky web they can leave many scales behind and still escape.

The scales carry the colours of moths and the colour patterns on the wings that are crucial for camouflage, deception, mate recognition and even potential predator communication. Moths, much more than any other insect group, have intricate patterns on their wings that have, no doubt, contributed to the success of moths. But it is not known if the development of scales coupled with the very general cryptic life strategy of moths has facilitated this. Scales may also improve the aerodynamic quality of the wings. Many very small moths have hindwings with very long scales on the trailing edge, said to be much more efficient than a broader wing.

Moth larvae are usually very generous users of silk. Silk helps them to cling to branches or to construct shelters for protection among leaves or to block the entrance to holes in branches. Tiny larvae float in the wind with the aid of a silken thread to find distant habitats, a means of dispersal important for moths with adult females unable to fly through loss or reduction of their wings and some others that are poor fliers. Silk can also be used for escape where a disturbed larva drops on a silk thread from leaves but is able later to regain its original position.

Finally, the development of the adult proboscis has permitted the exploitation of rare, high-energy food sources, such as nectar, sap flows, fruit and the sugary secretions of sap-sucking bugs, which are important for flight. It permits the drinking of dew, and in rare cases of tears and blood.

Butterflies may also drink sweat, liquids from decomposing bodies and ion-rich water from soaks.

Why are there so many different moths?

Insects and other animals often live in more or less specialised niches. Increased specialisation enables a species to adapt to, and very efficiently exploit, a small niche. On the other hand, the advantage of being a generalist is that a much larger population may be maintained through the use of a wide variety of resources. Specialists may be less vulnerable to predation and generalists less vulnerable to extinction. So, in stable habitats many moths will become very specialised.

Very small animals can inhabit very small niches and there are very many more small niches than large ones. As the

Rainforest on the Atherton Tableland, Queensland. This is a community with innumerable microhabitats for moths.

A mallee habitat near Coolgardie, Western Australia. Arid communities in Australia have an amazingly diverse moth fauna. Photo: Murray Upton

complexity of an ecosystem increases there will be more and more different small niches and many of these will be quite rare. The problem with exploiting rare small niches is that, relative to the size of the niche, they are widely separated. So, a very small animal that can live sedentarily in a small niche, but which can also be highly mobile and so move from niche to niche to disperse with each new generation, as well as being highly efficient at finding a mate, has a great many niches to which it could adapt. Moths (and other insects) have taken advantage of the vast numbers of rare small niches and this is an important reason why there are so many species. The features making this possible are a complete metamorphosis, small size, an efficient mate-locating method and the great mobility conferred by flight.

How well do moths see?

Moth vision is completely different to human vision. Moths have large compound eyes containing hundreds of facets. Each facet is a light focusing and detecting organ. The compound eye permits all-around vision and is also very sensitive to movement because a moving object must move across the face of the eye, from one group of facets to another, which makes it easily detectible.

Some butterflies have very good visual acuity but there is a balance between visual acuity and light sensitivity, so nocturnal moths will not have as good a vision as butterflies and day-flying (diurnal) moths. Some butterflies have the widest spectral range of any animals, with vision extending from red (700 nanometres) to ultraviolet (300 nanometres). Most moths see from red to ultraviolet although not quite as far into the red or ultraviolet as some butterflies. This is greater than the human range, which does not extend into ultraviolet.

Many moths have three visual pigments (green, blue, ultraviolet) but some have four and some butterflies have five (red, green, blue, violet, ultraviolet). In the compound eye there may be different facets with receptors for green and blue or for green and ultraviolet. Moths are good at colour discrimination and even at very low light levels, such as on a star-lit night, they can distinguish colours, including ultraviolet, reliably.

Moths can also detect the polarisation of light. In polarised light the plane of the light waves are aligned, and this can help moths distinguish reflected or scattered light from direct light. It allows, for example, a honeybee to see where the sun is on a cloudy day. Blue receptors are sensitive to vertically polarised light and green to horizontal or obliquely polarised light; polarisation detection is not independent of colour.

Why are moths attracted to light?

At night moths are most attracted to light in the near ultraviolet (wavelength 360 nanometres), which is just beyond the range of human vision. Astronomers say there is very little ultraviolet light available at night, yet moths can still detect this. It has been shown that moths are more strongly attracted to a point source of light than to a panel of the same luminosity. It has also been suggested that it is not the light source itself that they fly towards but the point of highest contrast between light and dark.

Anyone who has ever flown an aircraft knows that pilots orient themselves on the horizon. If the horizon is lost, as by flying into a cloud, then without instruments, the pilot rapidly becomes disoriented. Possibly moths may also orient on the horizon when in flight, at least in open situations.

Except for the face of the moon, the brightest part of the night sky is said to be just above the horizon and if this is true in ultraviolet then moths could orient themselves by viewing the ultraviolet light just above the horizon and the horizon itself will be the point of highest contrast. On cloudy nights the horizon may still be visible to moths because of their ability to detect polarised light. If moths do orient themselves on the horizon then a bright light will distort the apparent horizon into a circle around the light, which would explain their attraction to it.

Many night-active moths may come to light only rarely. Some will not approach

Many moths may be attracted to a bluish light on a favourable night.

very bright lights while others will ignore dim lights. Moths often prefer to fly at a particular time of night and this can depend on the conditions, particularly the temperature. Very active moths are more likely to come to light. Moths are extremely sensitive to the weather, responding to factors such as temperature, humidity, wind speed, moonlight and perhaps air pressure. The warmer the better, the more humid the better, and moths fly freely in rain provided it is warm.

Moonlight affects the numbers of moths attracted to lights in two ways. There seems to be a decrease in activity with bright moonlight. Moonlight also works as competition for the attracting light as there is approximately 100 times as much ambient light at the full moon compared to the new moon. The position of the light is important: lights looking down a slope may be more attractive either because the moths prefer to fly up a slope or because on a steep slope the light is closer to the canopy of the trees growing down the slope.

Cold air lakes and drains at night are usually too subtle for humans to notice but they make a big difference to moths. Moths will always try to land on the sheltered side of an object. They are extremely sensitive to wind; they can detect airflows that humans cannot and the slightest current will influence them. More importantly, tiny moths with big wings and low wing loadings must land upwind as aircraft have to in a relatively rapid airflow.

How do moths find mates?

Finding mates is achieved so efficiently in both butterflies and moths that the vast majority of specimens found in nature will have already mated. Nocturnal moths rely on pheromones (chemicals produced to communicate with other members of the same species) and use these very efficiently. Pheromones are carried on a plume downwind rather than disseminated in all directions. Almost universally, the female produces the pheromones from a gland on the abdomen but an exception is found in a few hepialids where the males produce a pheromone with a strong fruity fragrance. It makes sense that the females produce the pheromones while they rest quietly in greater safety while the males, actively patrolling the air currents, are more exposed to predation.

Moths use their antennae to detect the pheromones. A male emperor moth is able to detect a single pheromone molecule while several molecules are enough to elicit a response, and he can be attracted from a kilometre or so downwind of a female. The efficiency of pheromone detection has permitted winglessness to develop numerous times in many families leading to such curiosities as the psychid females that never leave their pupal cases or cocoons.

There is a very clear distinction between the diurnal moths and the nocturnal moths. The strictly diurnal moths behave like butterflies and recognise their mates visually from a distance. Butterflies use pheromones only when close to a mate as a check that it is the right sex and species. This difference is because moths cannot see as well at night as butterflies can in daylight, so visual recognition is not as reliable, nor can it operate over a useful distance. More importantly, during the daytime the sun heats the ground, causing local thermal air currents essentially moving up or down. Releasing pheromones into this disturbed air is inefficient. At night the air currents are

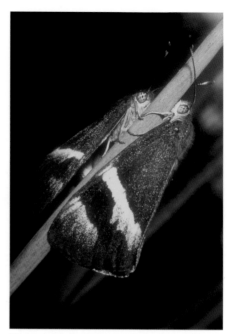

Well-developed pectinate antennae, such as those of *Laelia obsoleta* (Lymantriidae), occur in the males of many families of moths. Their antennae can process a lot of air when the moth is flying, thus helping them detect the pheromones produced by the females.

Mating sun moths, *Synemon directa* (Castniidae), rest in a characteristic pose for the family. In these day-flying moths mating may take several hours and the moths may stay coupled overnight. This is an uncommon form of coupling where the moths face the same way. Photo: Jan Taylor

calm, steady and parallel to the ground, making them much more effective in disseminating pheromones and a pheromone trail is easier to follow. The clear wing moths (family Sesiidae) are a notable exception; they release pheromones in the daytime.

The wide use of pheromones is one reason why pectinate (feathery) antennae that contact a much greater volume of air than simple ones are found widely in moths.

Overseas collectors often rear a virgin female from a wild-found larva and then use the freshly emerged adult female to produce pheromones that attract wild males. One male is allowed to mate with her and a whole new generation can be reared from the eggs. Using a captive virgin female in this way is called 'assembling'.

Some moths and butterflies remain near the foodplants to meet mates. A notable strategy in butterflies is hilltopping, where males of species with sparsely distributed individuals will congregate on hilltops or other prominent peaks like high trees. The females go to these places to find mates but this behaviour has not been investigated in Australian moths, although it is evident in many day-flying agaristine Noctuidae.

This is a mating pair of *Amata bicolor* (Arctiidae), male on left, showing the most common form of coupling where the moths face in opposite directions. The moths are from the Cairns district, Queensland, and probably mimic wasps.

Can moths hear?

The small moths (micros or microlepidoptera) with a few exceptions do not hear, but many of the larger moths (macros or macrolepidoptera) can. Moths hear with ears (tympanal organs) that may be at the rear of the thorax or at the base of the abdomen. Moths of most of the more modern families have ears and these are important in classification because there are often differences in the position and structure of their ears. Unfortunately, they are well covered in scales, easily mistaken for the counter tympanal hood (an ear-like structure to collect sound equivalent to the human external ear) and difficult to observe without destroying parts of the moth. The noctuids and arctiids have ears at the rear of the thorax while the pyralids, geometrids and drepanids have ears at the base of the abdomen.

Moths can hear in ultrasound and their ears are adapted to hear the calls of bats hunting by sonar. When a moth hears a bat it takes avoiding action. As a bat approaches a moth the rapidity of the bat's call increases so that it can get a clearer picture of the target and the moth determines from the rapidity of the calls when to duck, dodge and weave. Some moths can create their own ultrasonic calls and the arctiids have a structure on the thorax that they can flex rapidly to produce a call. Many species of arctiids can 'talk to the bat' or at least the bat recognises the call of these moths as an indication that they are inedible. Bats that

The ribbed sound-producing organ on the forewing of *Hecatesia exultans* (Noctuidae). The knobs on both forewings strike when the wings meet overhead causing the ribbed membrane to flex, making a whistling noise. Photo: John Green

have experienced a bitter tasting moth will leave those with a similar vocabulary alone. Not all arctiids are poisonous or taste bad and some may jam the bat's sonar with a very rapid burst of calls at the same frequency. It may be no accident that the more modern families, which are the most diversified and species-rich, the noctuids, pyralids and geometrids, have ears. It is thought that the ability to hear bats and so avoid bat predation has allowed the proliferation of the large modern moth families that may fly freely beyond the shelter of dense vegetation.

A few moths have used their hearing ability to communicate between the sexes. The agaristine noctuids, which are mostly diurnal, include some species where the males have an elaborate ribbed structure on the wings with which they make a 'whistling' or clicking sound. In Australia, *Hecatesia*, the whistling moths, are well known and *H. fenestrata* males fly a figure-of-eight course at dusk, whistling to attract females. *Hecatesia exultans* perches to call.

A genus of pyralid moths, *Syntonarcha* (see p. 135), also found in Australia, has the last segments of the abdomen and male genitalia modified to produce a scraper and a ridged file that it uses to call in ultrasound to attract mates.

There is a group of mites that infests the ears of noctuid moths including the Australian moth *Mythimna convecta,* of which about 2% of the Canberra population is infested in summer. These mites can build up in very large numbers in the ear and affect hearing. When one of these mites crawls onto a moth it first checks both ears and if one is already infested it will settle down in that ear. The mites always leave one ear uninfected. It is no use to either the mites or the moth if the moth is deafened by having both ears infected and then eaten by a bat and this elaborate behaviour by the mites is an indication of the intense selection pressure due to predation from bats.

Why are there so many wing shapes, patterns and colours?

The variety of colours and patterns, shapes and sizes in moths are not for human benefit but are for the benefit of the moth itself. They may serve to deter predators or influence other moths of the same species, or they may provide some physical benefit. For example, hawk moths have a high wing-beat frequency and have developed narrower wings; other moths have a reduced wing size because of a sedentary habit or a response to the dangers of being blown away in windswept but limited habitats.

It is also said that narrow wings create less drag and that in very small moths a trailing edge of the hindwing formed of long hair-like scales has an aerodynamic advantage. But the real reason for the bizarre wing shapes shown by pterophorids and alucitids is not known. Other physical benefits may entail thermoregulation if the species is diurnal and basks to warm up. In this case black absorbs heat more rapidly.

The wings of most moths are coloured and patterned to camouflage the resting moth. For example, moths such as *Hednota aurantiaca* (p. 134) that live in sedges or grasses have a prominent pattern of parallel longitudinal lines along their wings. Even more common is a transverse pattern of bands that helps break up the long outline of the wing.

Such a pattern of bands usually meets the leading edge of the forewing at right angles, and, for aerodynamic reasons, the leading edge needs to be straight or gently curved. Straight lines are rare in nature and can aid a predator to recognise a moth and so the lines may be disguised by scalloping, and dots, spots and banding with the bands usually curved, zigzagged or broken.

One very common pattern repeated in many Australian families is a golden wing with a broad tip of brown (for example, *Chrysonoma* sp., p. 65). Gold is a very

The larva of *Phyllodes imperialis* (Noctuidae) rears up when threatened to resemble a reptile. This larva feeds on the vine *Carronia multisepalea* (Menispermaceae) and belongs to an isolated population in the Brisbane area. Photo: Don Sands

This giant moth, *Phyllodes imperialis* (Noctuidae), closely resembles a dead leaf but the hindwing, hidden here, is brightly coloured. The moth, whose slow-moving larva looks like an active reptile, is found in New Guinea, the Solomon Islands, New Hebrides and Australia.

15

This moth of the genus *Eucymatoge* (Geometridae) shows the multiple transverse banding on the wings, which breaks up the outline of the moth, making it difficult to distinguish.

prominent colour in dry Australian grassy woodlands and the brown tip may serve not only to break the outline of the wing but to make the moth look smaller and less attractive. But there must be more to it than this as the pattern is also common in rainforests.

Camouflage can be sophisticated, as with the geometrids and noctuids that rest with wings flattened closely to the substrate and appear just like the lichen or leaves they rest on—even down to tufts of scales to give the deception some depth. These moths are camouflaged to look like part of the substrate upon which they rest. *Phyllodes imperialis* adults look like a dead leaf but the larva takes a more active pose, attempting to look like a reptile.

Some notodontids and the xyloryctids (*Cryptophasa xylomima,* p. 77) greatly resemble a broken piece of rotten wood.

Some may look like dead or live leaves, stems, buds, shoots, bark or lichen. Others may look like a bunch of hanging dead leaves, which is probably what the Hercules moth in its resting position looks like—it even has clear windows where the light may shine through. Overseas it has been shown that some moths that rest on the ground differ in colour with different soil colours. This has not been demonstrated in Australia but it probably happens here also. In rainforests or other areas with low light levels, nearly transparent wings may help a moth blend with its background (*Didymostoma aurotinctalis,* p. 129).

Some moths may look like the more repulsive or threatening parts of the environment. They may look like bird droppings, thorns or even holes in dead leaves (*Hoplomorpha abalienella,* p. 75). The

This moth, *Epicyrtica metallica* (Noctuidae), is well hidden among the lichens upon which it is resting.

resting hepialid *Zelotypia stacyi* is reported as resembling the head of a goanna (p. 42). One geometrid, *Pingasa blanda*, (p. 144) has a wing pattern that may suggest a hunting spider.

Other moths, particularly saturniids and noctuids, practise a more active defence and have fake, large, mammal-like eyes on the hindwing that may be covered at rest but exposed when threatened by a predator. Such eye-spots are often not very good resemblances but they are enough to cause surprise and hesitation long enough for the moth to escape. Even flashes of bright colours exposed by moving the forewing forward may cause enough hesitation to allow escape. Some diurnal moths in the rainforest where light contrasts are exceptionally high are black and white—a common occurrence in rainforest butter-

flies. The flashing of black and white can cause confusion with bright dapples of light that also appear to move from parallax as the predator moves (*Abraxas expectata* p. 139).

Another common pattern in very small moths is a series of small black dots, often highlighted with shining metallic scales set around the outer edge of the wing (*Eupselia satrapella*, p. 78). The purpose of these is not known but most of the moths with it are diurnal and it has been suggested that the dots may look like the multiple ocelli of spiders.

Many moths contain poisons from the plants they eat as larvae and others produce poisons, such as cyanide, themselves (*Amerila rubripes*, p. 181). They can advertise their poisonous nature with bright colours and distinctive patterns so that only

This moth, *Argina astrea* (Arctiidae), is boldly coloured and flies lazily, sometimes in broad daylight, advertising its poisonous nature. The caterpillars feed on the pea plant *Crotalaria* and the moths are protected from predation by the pyrrolizidine alkaloid poisons contained in the plants. The moth is found from Africa to northern Australia.

naïve predators will try to eat them. Educating predators is easier and less destructive to a population if different species share the same colours and patterns. So, many different species that are all poisonous may look alike. The genera *Utetheisa* (p. 182) and *Argina* that contain at least four distinctly spotted Australian species are a good example. They contain pyrrolizidine alkaloids and the species of *Utetheisa* are all very similar, spotted in red and black on white (p. 182).

Some edible moths may take advantage of this and resemble poisonous ones, deriving benefit from the protection from predators conferred by their colours when they are in fact edible. Some moths that are not poisonous try to look like an animal that is dangerous to a predator. Sesiid moths

(*Ichneumenoptera chrysophanes*, p. 107–108) usually resemble wasps and species of the oecophorid genera *Snellenia* (p. 68) and *Pseudaegeria* also resemble wasps. Other insects like bugs and beetles may also be part of these 'mimicry rings'.

Patterns and colours may also be used by day-flying moths to distinguish the opposite sex and even by some night-flying species like the pyralid, *Hypargyria metalliferella*, and the noctuid, *Mythimna decisissima*, in which males have the underside of the wings a shining silver while the females are dull grey.

The larvae also have a very wide range of protective strategies. Larvae feeding or living in shelters are mostly plainly coloured, sparsely hairy and of normal shape. Exposed larvae may be very hairy

This moth, *Pseudaegeria* sp. (Oecophoridae), mimics a braconid wasp so that predators mistake it for a stinging wasp and leave it alone. This genus includes a suite of species that mimic distasteful beetles or wasps.

This is one of the wasp models in the family Braconidae that the *Pseudaegeria* moth is mimicking. The moth and bug shown mimic it very closely.

This is a bug of the family Miridae that also mimics the same braconid wasps as the *Pseudaegeria* moth. Mimicry rings involving many species, often of different orders, occasionally involve moths in Australia.

When disturbed, the moth *Opodiphthera eucalypti* (Saturniidae) throws its forewings forward suddenly revealing the eye-like rings on the hindwing. The forewing spots are small, as if closed, but the hindwing eyes are wide open and awake. No wonder predators are intimidated and think twice before attacking.

with hairs that are dense and long, sharp or urticating; they may also have spines, or even stinging spines. Presumably hairs and spines are uncomfortable in the throat of vertebrates. Some lasiocampids have hairs along their sides that extend down to the branch they rest on in order to soften their outline.

Some larvae may eject poisons from special glands. Others may resemble dangerous animals; some Notodontidae look a bit like scorpions. Such larvae may display when disturbed to look more like their models. They may have eye-spots to make them more lizard- or snake-like. They may contain poisons from the foodplant or make their own and may have bright colours to warn predators. Some large larvae have counter shading where the underside is paler than the upperside so as to nullify the visual effects of shadow if they rest right way up.

How long do moths live?

Adult moths can live anything from one day to perhaps eight or nine months and it depends very much on the life strategy that the moth has adopted. Species with non-feeding, highly active adults, for example the hepialids, live only a day or two. Moths like the bogong moth with a biology involving a long return migration will live for up to eight or nine months. However, the life expectancy of the general run of moths will be from one to three weeks.

The entire life cycle may take a much more variable time. Some fast-breeding moths in warm climates may take only a month or a bit more for their entire life cycle. But most are constrained by the climate and many moths even in warm climates have an annual life cycle. Some may fly in spring and autumn and have two

There are many insect predators of moths. Ants, in this case a bull ant, can be very destructive, attacking any of the stages of the life cycle.

This tachinid fly is attempting to lay eggs on a hairy caterpillar but the fly is having difficulty and the hairs are providing some protection. Tachinid eggs may be ingested with food or the flies may lay on the caterpillar's skin. When the maggot hatches it burrows into the larva and feeds internally.

cycles per year. Many breed rapidly over summer and with the onset of winter go into a long resting stage.

In Australia, it is often heat and dryness that determines the life cycle. A very common strategy in mallee and other seasonally dry habitats in southern Australia is for the adults to emerge in autumn (sometimes stimulated by autumn rains) and to lay eggs immediately. The eggs may hatch promptly or wait for further rain before hatching. The larvae feed over the winter period when it is moist and plants are growing and then burrow into the soil with the coming of hotter weather in September or October to pupate in the soil where they spend the whole summer before emerging in April or May.

A common strategy in rainforests in northern Australia, particularly the drier monsoon forests, is for the adult moth to emerge at the beginning of the wet season and lay eggs on the new shoots; the larvae feed while the plants are growing over the wet season and they pupate late in the wet

or after the wet is over. These moths then have a resting pupal or pre-pupal stage until the following wet.

The very large cossids may take two or three years to complete their life cycle with most of this time as larvae. Moths that take two years could theoretically separate into two parallel populations, one in even years and one in odd years. In practice, the period seems variable enough for this to not happen; at least, no case is known in Australia. Some overseas butterflies are known with a two-year life cycle with adults only in odd or even years where the populations have become synchronised, possibly by the elimination of one of two parallel populations. Similarly, in theory, a moth that flies in spring and autumn could have a six-month life cycle or it could be that two annual life cycles are living together. Again, this latter possibility is not known in Australia and again there is probably enough variability to prevent the development of two essentially different populations. A few moths have a very long resting

Even deep in timber the cossid larva is not safe. This ichneumonid wasp has an enormously long ovipositor designed to penetrate thick tree trunks in order to reach a hapless cossid larva.
Photo: Ray McInnes and John Green

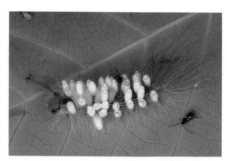

These braconid wasps, after eating out the insides of the larva, have emerged and spun white cocoons among the hairs. The adult wasps are now emerging. Parasites play a critical role in controlling moth numbers.

stage in the cocoon either as a pupa or pre-pupal larva that may extend for up to 12 years.

Moths are prey to many predators and diseases and are attacked at any stage of their life cycle. Birds, reptiles and bats are all important predators of adult moths. Hunting insects such as robber flies, praying mantises and ants may also eat them.

Web-building spiders trap some moths, as do the pouncing and ambushing spiders. Losses are greatest in the early stages of the life cycle with eggs and young larvae eaten by hunting spiders, sucking bugs and predatory wasps.

Most destructive of all are the parasites, which are mostly wasps and sometimes tachinid flies that attack the eggs or larvae but may emerge in a subsequent stage. Tachinid flies may lay their very small eggs on the skin of the larva where they hatch and bore into the larva, or they may lay on the foodplant and the larvae ingest the eggs while feeding. The parasitic wasps sting the egg or larva, depositing an egg in the process. The egg may develop normally or it may divide many times before starting to develop and so produce many identical larvae. When they complete development the wasp larvae may emerge and spin cocoons, or they may spin cocoons within the larva or pupa.

There is also a wide range of fungal, bacterial and viral diseases, usually attacking the larval stage. The so-called 'vegetable caterpillars' are well known: these are hepialid larvae that are riddled with the fungal hyphae of the *Cordyceps* fungus that then sends a fruiting body to the surface, like a mushroom. The bacterial disease *Bacillus thuringiensis* is widely used as an environmentally friendly insecticide. A suspension made from bacterial cells is sprayed on a crop to infect feeding larvae such as *Helicoverpa* spp. in a cotton crop. Such a spray leaves few chemical residues, although it may infect larvae other than the target pest.

Why are particular moths found in some places and not others?

Where each species of moth occurs depends on many factors that all interact in determining distribution. Some moths are only found in very small areas while others are very widely distributed.

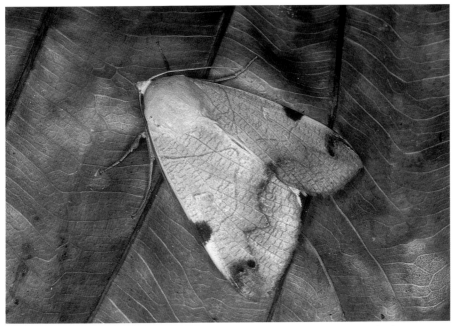

Some moths occur over very large areas of the globe but others may be restricted to small islands or mountaintops or rainforest patches. This moth, *Ophiusa tirhaca* (Noctuidae), is a very mobile and widespread moth known from Europe to the Pacific Islands. Not all widespread moths are large; some are very small and travel on wind currents.

At all stages of their life cycles, moths differ greatly in their climatic requirements; they differ in their tolerance of temperature, rainfall, humidity and many other climatic variables. Some larvae like those of the noctuid *Apina callisto*, which are exposed openly on the ground in heavy frost in winter, tolerate freezing temperatures that many moths could not survive. A larva of the anthelid, *Anthela ariprepes*, from arid western Queensland and New South Wales, pupated successfully in the glove box of a car parked in the sun, a situation that would have cooked many moths. Some noctuid moths visiting a light in tropical rainforest may shiver to warm up their flight muscles in a temperature of 30°C. In contrast, some scopariine pyralids fly within feet of snow patches on Mt Kosciuszko and bask in the sun. So moths vary enormously in their tolerance of the extremes of climate.

Perhaps the most important factor determining distribution is the larval foodplant. Caterpillars of some species feed on a single species of plant, and then perhaps only on flowers or fresh shoots; those of other species may feed on related groups of plants; yet others on a very wide range of plant material and a few are carnivorous.

The saturniid *Opodiphthera astrophela* is most frequently found in rainforest in southern Queensland and northern New South Wales where it feeds as larvae on a variety of trees including *Flindersia* (Rutaceae). *Opodiphthera astrophela* is also found in arid western Queensland and New

This small moth, *Plutella xylostella* (Plutellidae), is a pest of cabbages. It is one of the most widely distributed moths found from Scandinavia to the subantarctic islands, Europe, Africa, Asia, Australia and the Americas.

South Wales in the driest habitats where the leopard wood (a species of *Flindersia*) grows. The moth is never found in the eucalypt forests distributed between the rainforest and arid zone and it is clear that the presence of a suitable foodplant determines its distribution.

Sometimes there is a more complicated foodplant relationship; there is a suite of moth larvae that feed only on the rust galls on *Acacia*. The relationship may involve other predators or parasites, or animals upon which the moths depend. For example, the pyralid *Ecnomoneura sphaerotropha* lives in bloodwood apples—tennis-ball-sized galls on *Corymbia terminalis*—caused by sap sucking insects. In some cases, females may be able to lay more eggs if adequate nectar sources are available and the species may survive only where these are available.

A moth that is found in rainforest may be dependent on the climate and microclimate in the rainforest or it may be there because its foodplant is there. Most commonly, however, the moth will be depending on both the foodplant and the climate and indeed the foodplant is probably also dependent on the climate.

Some moths only occupy part of the range of apparently suitable climate or suitable foodplants and often historical factors determine their distribution. Many moths are present in the Snowy Mountains of New South Wales but are absent from the correct foodplant in a similar climate in Tasmania, but these similar places have been separated for a long period by unsuitable habitats.

Most moths that are not long distance migrants are confined to the habitat in which they live and habitats can be very patchy. Alpine habitats might be confined to discrete mountaintops and, particularly in Australia rainforests or monsoon forests, tend to be very patchy. Where these patches have been continuous in the recent geological past, they tend to be inhabited by the same fauna. But if separated for long periods the faunas show differences. The rainforest fauna near Brisbane has many differences to that occurring around Cairns, which is itself different to that at Iron Range, about two-thirds of the way up Cape York Peninsula. The fauna at Iron Range is in many respects more similar to a depauperate New Guinea fauna than to that at Cairns.

The fauna of the Snowy Mountains is generally similar to that in Tasmania but species that are present only above 1200 m in the Snowy Mountains are often absent from Tasmania, but may be represented there by closely related species. The same is true of Tasmanian species only found above 1000 m in Tasmania that are usually absent from the Snowy Mountains. Why this should be is not known but it has something to do with the history of Australia during the ice ages.

The really big historical events for the Australian fauna were the separation of Australia from the rest of Gondwana and the increasingly close contact of the northern coast to Indonesia and New Guinea. Added to this was the extensive drying out of the continent during the Miocene geological epoch. It is suspected that many moth groups have a Gondwanan origin, for example the Agathiphagidae and Castniidae and tribes or subfamilies within many other families. There are also many families with massive radiations of genera and species in southern Australia that are clearly not related in any way to the Indonesian or New Guinean fauna, but nor is there evidence that they are Gondwanan. These groups are well represented in both sclerophyll habitats and in rainforests.

The most primitive moth group, the micropterigids, is as old as the dinosaurs and occurs almost worldwide, but in Australia is found only in rainforests. The family is found in New Caledonia (a chip from Gondwana) but not in New Guinea, which is, by geological standards, of recent origin. Other groups of mostly rainforest species are clearly recent arrivals from New Guinea, for example the noctuid, *Pterogonia cardinalis*, a conspicuous large moth with a single silver spangle on the forewing. It is a New Guinean species known only from Iron Range in Australia until recently, when it appeared around Cairns, showing that the southward spread of this New Guinea element is continuing.

In very recent times populations have been fragmented by habitat destruction due to agriculture. This affects distribution on a smaller scale.

New Zealand lacks many families present in Australia and has a fauna with some similarities and many differences. Most of the similarities can be explained by dispersal over water. One curious exception is the apparently close relationship between the New Zealand arctiid genus *Metacrias* and the Australian *Phaos*. In Australia, *Phaos* is found in the Snowy Mountains and Tasmania and *Metacrias* is found only on the South Island of New Zealand. Dispersal will not explain the distribution because they have flightless females and, so far as is known, no means of dispersing as tiny larvae. This and many other New Zealand groups, which are not represented in

Australia, have presumably been in New Zealand since it split from Gondwana. In another example, some New Zealand butterflies are placed in *Lycaena* (Lycaenidae); they have no relatives in Australia or South America but are found in New Guinea and are very widespread in Asia. Possibly they were in Australia at one time but the aridification of Australia during the Miocene eliminated them.

Have any moths become extinct?

No Australian moth is considered extinct but this is because so little is known about the distribution, ecology and identification of moths. The fact that someone may have collected a species at Broken Hill in 1900 which has not been seen since does not mean that it is extinct but it does mean that no one who could recognise it has looked for it since in the right place at the right time.

We know that some moths have become extinct over most of their range but still occur in small, protected sites. Several moths have been found only on Black Mountain, in suburban Canberra, although there is no proof that they do not occur elsewhere. The moths most vulnerable are those that originally had a very restricted natural distribution and it may be that many have become extinct before they could have been discovered. Many may have become extinct during the time of massive vegetation degradation following the development of inland New South Wales and Victoria for grazing between 1840 and 1910.

Many moths that still survive have been affected by the fragmentation of habitats making them even more vulnerable to wild fire, injudicious burning and to further land clearing.

Some moths have been listed as endangered or threatened where they are known to be confined to an endangered habitat. In Canberra in the 1990s the endangered castniid moth, *Synemon plana* (p. 104), was listed because the lowland native grasslands where it occurs were themselves listed as an endangered habitat. In cases like this there is usually a suite of species confined to the habitat, making a stronger case for protection. The lowland grasslands of the Australian Capital Territory contained an endangered legless lizard, an earless dragon and the mouthless moth. These all became 'figure head' species representing unknown others that were also conserved by the reservation of their habitat. Where enough is known about a moth it can contribute significantly to conservation in this way.

In conserving moth species it is important to investigate the processes that may lead to extinction. These are called 'threatening processes' and they cause habitat changes over large or significant areas. Many threatening processes have been identified, including extensive grazing, land clearing, clear felling, strip mining, housing development, weed infestation and control burning. The effects of control burning have received insufficient study (see p. 75) and because it is practised over large areas it is of particular concern.

Some threatening processes affect particular habitats, such as beach mining, marina development and climatic warming, and the latter has the potential to eliminate many cold-adapted species from isolated mountaintops.

Collecting moths is the only way to acquire reliable information about them. No butterfly or moth in the world has become extinct through overcollecting

This female castniid, *Synemon selene*, is a large day-flying moth basking in its native grassland habitat. It is from a western Victorian population thought to be parthenogenetic (females lay fertile eggs without mating) and lacking males. The only known bisexual population near Adelaide may be extinct. Photo: Fabian Douglas

except for a few cases where habitat destruction had decimated populations to the point where survival was unlikely anyway. The extinction of the large copper butterfly in Britain is a classic example in which heavy collecting coincided with massive drainage of the fens leaving little suitable habitat. Because collecting is necessary for our knowledge of moths, the provisions of most endangered species legislation in Australia are a mixed blessing.

Legislation provides some protection through requiring impact statements, helping restrict development of sensitive areas, and helping to fund research into endangered species. But in restricting collection it also restricts the acquisition of further knowledge.

What use are moths?

Moths exist in large numbers because, in evolutionary terms, they have been very successful. Considered from this viewpoint no animals have a 'use' as evolution has no direction or aim. People who ask this question mean, 'What use are moths to me?'. The obvious direct way moths have been used is in the production of silk. The silkworm *Bombyx mori* (Bombycidae) has been domesticated for thousands of years. There are also several well-known wild species in the family Saturniidae that have been used for silk production: Tussah silk is produced in India from the cocoons of *Antheraea paphia*, Muga silk from *Antheraea assamensis*, Tensan silk in Japan from *Antheraea yamamai* and Eri silk from *Samia cynthia*.

This is the adult moth of *Cactoblastis cactorum* (Pyralidae), the famous cactoblastis moth that controlled the serious weed prickly pear in southern Queensland in the late 1920s and early 1930s. Prickly pear is still present and without the continued control exerted by the moth would become a serious pest again. Photo: Ian Common

This is the larva of *Cactoblastis cactorum* feeding on prickly pear. Photo: Ian Common

Altogether, about 20 species in the families Lasiocampidae, Saturniidae and Notodontidae have been used by native peoples in addition to the silkworm. A webbing sheet covering the cocoons of *Opodiphthera sciron*, one of the saturniids, in New Guinea was used as a waterproof head covering by the New Guineans.

More indirectly, moths have been used to control weeds: the classic example of this is the control of prickly pear in Queensland and northern New South Wales in the late 1920s and early 1930s by the pyralid, *Cactoblastis cactorum*. The prickly pear is still there—it has not been eliminated but controlled—and the moth, though now rare, continues to prevent prickly pear from breaking out again. The beauty of it is that, except for a small initial investment, this control costs nothing.

Moths are still tested for suitability and imported for the control of weeds and, in many cases, probably exert an effect even if not as spectacular.

Most important, however, is the fact that moths are an integral part of the natural and agricultural ecosystems by which we live. Eliminate the moths in some thought experiment and these ecosystems would be changed completely. Much has been said on the 'arms race' between plants and herbivores with each trying to gain an advantage. Usually this is seen in terms of the development of plant protective chemicals to discourage or poison insects although it involves many other strategies as well. There would probably have been no poppies or pot without the herbivores to poison (if you are interested in such things).

Many pharmaceuticals owe their discovery, if not their production, to poisons found naturally in plants. Quinine is used for malaria; morphine and derivatives of cocaine as painkillers, as well as drugs from the tomato plant family for eye surgery and at one time for motion sickness. Other plant groups that have strong poisons are the asclepiads brimming with heart poisons and all the borages equally replete with liver poisons.

In Australia, the Myrtaceae are full of oils, not only to burn competing plants in bushfires, but to make life difficult for insects as well. Many plants, from grazing sorghums to cherry laurels, produce cyanide, a poison so effective that the insects have taken it up themselves, as they have some of the other chemicals mentioned. In a roundabout way we owe all these pharmaceuticals to the work of herbivores—including moths. Remove the herbivores (remembering that moths are the leaf eaters *par excellence*), and the world would change.

More obvious ecological services performed by moths include pollination (p. 166), recycling of organic matter (p. 74), and, of course, they are a source of food for birds and small vertebrates. A few moth larvae attack scale insects and so may be beneficial in helping to control these:

Stathmopoda melanochra (p. 65) in the family Oecophoridae, *Creobota coccophthora* in the Pyralidae, and *Metaeomera* spp. in the Noctuidae all help to control outbreaks of scale insects.

It has been said that a very high diversity helps to keep ecosystems stable and that their simplification for agriculture is apt to produce plagues and perturbations. On this thesis every moth, no matter how obscure its activities, is helping to keep this planet stable in however small a way.

Moths have contributed greatly to scientific research with their study helping our understanding of evolutionary processes, with the work on industrial melanism, and how strong natural selection pressures can be in nature some of the best-known examples. Studies of the genetic control of polymorphism (the many forms of mimetic species, for example) led to our understanding of human blood groups.

Moth metamorphosis has been used as a paradigm for people dealing with death. Moths or butterflies have been used in art since the Egyptians and still provide inspiration and wonder. The beauty and diversity of moths can instill a sense of wonder and respect for our natural environment that, for the sake of our future, many people in positions of power would do well to cultivate.

Why study moths?

Detailed studies of pest species are necessary if they are to be controlled efficiently. The more that is known about the biology of a pest then the more options there are for control methods that do not involve chemical poisons, can be more precisely targeted to a particular pest, and are more environmentally friendly. Such studies require biological research on the pests and also

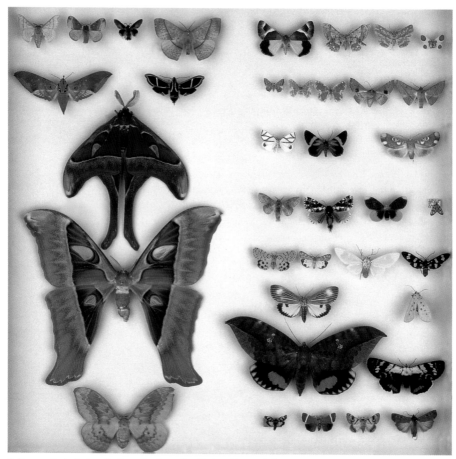

A drawer of selected moths from the Australian National Insect Collection, CSIRO. Identification of moths is an essential first step to learning and communicating about them. Photo: David McClenaghan

require research enabling a particular pest to be distinguished reliably from other pests and innocuous species.

The most serious moth pests of crops in Australia are two very similar species of *Helicoverpa* (Noctuidae) (see p. 194). Methods found in the 1950s to distinguish these and similar species still underpin all research into the control of these moths. Even where insecticides are used that may be effective against two very similar pests, insecticide resistance will develop inde-

pendently in each and there will also be differences in biology.

Quarantine measures are in place to keep many pests out of Australia. To establish these measures we need to know what moths are already naturally found in Australia as well as which pests we want to keep out. We must be able to recognise all these species, in their various life cycle stages, and know how they are most likely to enter the country. The Asian gypsy moth has caused great damage to forests in the United States and it is very desirable to keep it out of Australia, so protocols have been formulated to permit its recognition at possible entry points.

Some moths that are pests overseas are not pests in Australia even though they are present. It may be that there are cultural differences in growing the crops between Australia and, for example, Indonesia, or it may be that there are more effective parasites and predators in Australia. It is important to recognise these potential pests because changes in conditions and agricultural practices may lead to them becoming more serious pests here. Recently, a second sugarcane army worm was recognised from Australia and this second species is the more serious pest overseas. However, neither has been a very important pest in Australia and we know that the most recently recognised species has been in Australia for a very long time because old specimens are present in collections.

Foreign insects imported for the biological control of weeds must be thoroughly tested in the laboratory before release to ensure that they do not attack plants other than the target weed. It is essential that only one species is imported and that it is correctly identified as the most promising species. Where wasps or flies are introduced to control moth pests, then a detailed knowledge of the relatedness of various pest and innocuous species is necessary to systematically determine what the host range of the wasp may be if it were to be released in Australia.

In conservation it is also important to accurately distinguish species. An endangered or threatened species can only be listed if it is clearly distinguished from other similar species.

Knowing what moths contribute and what they do is important in understanding

A drawer of very large hawk moths, *Coequosa triangularis*, in the Australian National Insect Collection, CSIRO. Photo: John Green

the natural functions of moths in ecosystems, for example, pollination, nutrient recycling, food for other organisms, and control of leaf litter and plants, to mention only a few.

To study moths it is essential to have a single internationally agreed name for each moth. The system that we use is the closest thing there is to an international language. A name is essential—without it no knowledge can be accumulated or communicated about moths. Try an experiment: take any sentence in this book and cross out all the nouns (naming words) and see how much sense it makes. We know our own names but the insects do not know theirs, which means there has to be someone who does know the insects' names and is able to recognise new ones and give them a name so that any other biologists and the general public can communicate about them. We need people who can identify moths and know about their biology and behaviour. In Australia the number of such people is dangerously few.

This unidentified gracillariid from Renmark, South Australia, has emerged from a gall made by the larva in a shoot of *Eremophila oppositifolia* (Myoporaceae). Photo: Marcus Matthews

IDENTIFYING THE MOTH FAMILIES

Many of the moths seen at the light in the porch or at the window can be identified to family from this book. However, it will not always be easy, particularly with the smaller moths. Wherever possible, characteristic features that can be seen without a 10× lens and can be seen on a live moth have been used, but often other features that need a dead moth have to be used as well. These features are listed for each family in a bulleted series. Some more distinctive features may be mentioned, but need closer examination with a 10× lens. Sometimes features that need dissection or a microscope are mentioned so that the differences in the families can be appreciated, although the general reader will not be able to see the features.

The moths are a very large and successful group commonly supposed to have speciated rapidly with the rise of the flowering plants. Their rapid evolution means that many features that seem distinctive are not useful in identifying them because there are too many exceptions. When identifying a moth look for *all* the features listed any one may be misleading.

Those familiar with biological keys to identification will wonder why keys are not used. Keys to Lepidoptera are difficult to use as the most important features in their classification require the specimen to be destroyed to see them and if broken up to see one feature then others will be lost. Moths are covered in scales making structures difficult to see, and the very useful structures of the genital system are with-

drawn into the abdomen. Most lepidopterists use something like the list of features given here and, when all agree, then the family can be identified.

Where the reader can match a photograph with a moth then a species identification of that moth is possible because the colour pattern of the wings is often distinctive.

There are some very large families and the species from these are the most likely to be encountered. If the moth is small, check the Oecophoridae and Tortricidae first. If it is large, then check the Pyralidae, Geometridae and Noctuidae. These five families contain the vast majority of Australian moths.

There are 85 families of moths in Australia but the status of some is controversial and several recent changes have been suggested. The 69 families treated here are not an up-to-date list but one that is consistent, but not identical, with recent books on Australian moths. The families are arranged in a sequence commencing with those that retain the more primitive features and progressing to those with more recently developed features.

Some very rarely seen families, those containing only one species, or, for which we lacked a living photograph, have been omitted. These are: Lophocoronidae, Palaeosetidae, Anomosetidae, Incurvariidae, Eriocottidae, Arrhenophanidae, Douglasiidae, Heliodinidae, Lyonetiidae, Batrachedridae, Coleophoridae, Blastodacnidae, Agonoxenidae, Momphidae, Macropiratidae and Simaethistidae—most of them tiny moths.

The manner in which the forewings and hindwings are coordinated in flight is one of the best ways to distinguish males from females. In male moths, on the underside of the wing a strong bristle (frenulum) from the base of the hindwing slips into a hook-like flap (retinaculum) of cuticle on the underside of the forewing.

In females there are almost always several bristles forming the frenulum and a patch of raised scales representing the retinaculum instead of a hook. The frenulum and retinaculum may be absent, as in many bombycoids, where the wings work together because of a generous overlap between them. The frenulum and retinaculum are also absent in the most primitive families. The frenulum is hard to see in living moths and even in dead and spread individuals one mostly needs a 10× lens.

In a progression from the most primitive families to the most modern there are many features that have changed. Looking at these features can place families in an evolutionary sequence. (Primitive families are those that currently retain a high proportion of features associated with Lepidoptera as they were millions of years ago when they first evolved, but they may still be highly evolved in other ways and, of course, all living organisms have evolved over the same period of time and none are more ancient than others.)

The sequence of families proceeds from those that have retained many of the putative primitive features to those with few. There are many features that have changed but, unfortunately, most need magnification to see but they themselves are interesting as they show how moths evolved.

The families Micropterigidae and

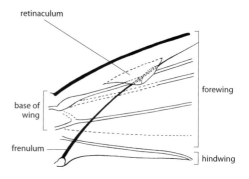

Fig. 1. The base of the underside of the wings with scales removed, showing the frenulum and retinaculum of a male moth. B. Rankin

Agathiphagidae have chewing mouthparts as adults. The Hepialidae have a proboscis with muscles in it as do all the families after them unless it has been secondarily lost (as often happens). The families before Nepticulidae have the forewings and hindwings of similar shape and venation; Nepticulidae and those after have hindwings with reduced radial veins, which are thus often smaller. The families before Psychidae have a single copulatory and egg-laying opening in the females; Psychidae and those after have two separate openings with an internal connection so that sperm may be transferred from the copulatory to the egg-laying organs. The Hepialidae (and some families omitted) have an external groove that extends between the two openings. This is a curious parallel to the monotreme distinction in mammals, monotremes with one opening and the rest of the mammals with two. The median vein (Fig. 2) is present and tubular within the

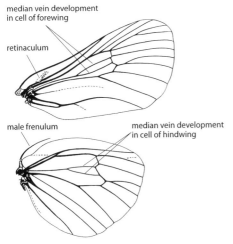

median vein development
in cell of forewing

retinaculum

male frenulum

median vein development
in cell of hindwing

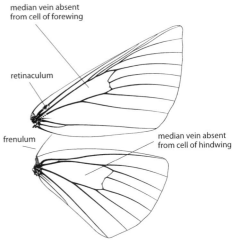

median vein absent
from cell of forewing

retinaculum

frenulum

median vein absent
from cell of hindwing

Fig. 2. The veins of the wings of a psychid moth, with scales removed, showing the median veins present in the cells of both wings. B. Rankin

Fig. 3. The veins of the wings of *Carthaea*, with scales removed, showing the absence of the median veins in the cells of both wings. J. Wedgbrow

cell of one of the wings (the cell is the area near the centre of the wing that is encircled by veins) in the families up to and including the Cyclotornidae (except for all the gelechioid families).

Up to and including the Cyclotornidae, except for the gelechioid families, the pupa is mobile and because it has spines on its upper surface as it squirms it is able to push itself forward. The other exception is the Pyralidae, which have a mobile pupa but are placed after some families without. The mobile pupa means that the pupa is extruded from the cocoon before the moth emerges, whereas if the pupa is less mobile then the adult emerges from the pupal skin and then forces its way out of the cocoon, which is quite a fundamental difference.

All the major families before and including the Pyralidae feed in shelters of silk, or in hidden situations, while the families after feed openly on the vegetation, even if only at night in some cases. The more

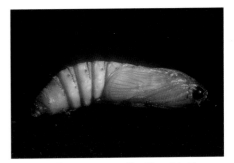

This pupa of *Synemon magnifica*, (Castniidae) has short spines on the abdomen that, when the body is wriggled, push the pupa forward. The pupae of these moths protrude from the cocoon or shelter for the moth to emerge. Photo: Ted Edwards

primitive families have conspicuous maxillary palpi next to the proboscis as well as the prominent labial palpi. These are the clearly visible scaled minute 'fingers' protruding in front of the head (Fig. 4); the maxillary palpi are usually smaller and more hidden.

The maxillary palpi are sometimes clearly visible in the Pyralidae but they are

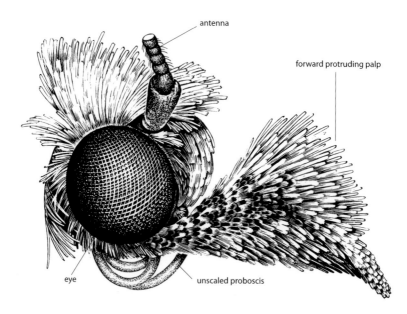

antenna

forward protruding palp

eye

unscaled proboscis

Fig. 4. The head of a moth, family Carposinidae, shows the naked proboscis and the forward pointing labial palpi. F. Nanninga

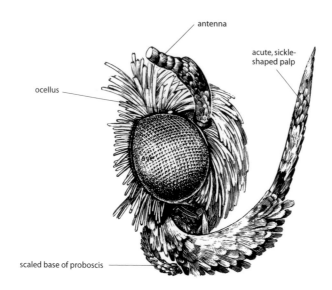

antenna

acute, sickle-shaped palp

ocellus

eye

scaled base of proboscis

Fig. 5. The head of a moth, family Oecophoridae, shows the scaled base of the proboscis and the sharp-pointed, up-curved, sickle-shaped labial palpi. F. Nanninga

not usually prominent in families after the Tineidae. They are absent or minute in families beyond the Pyralidae. With a few exceptions, the Pyralidae and the families after them have hearing organs allowing them to detect bats. These have arisen several times independently—those on the Geometridae and Pyralidae are on the abdomen, those of the Noctuidae and Arctiidae are on the thorax—but they are fundamentally different. It is probably significant that, except for small moths, the very species-rich Lepidoptera families (those with over 10 000 species worldwide) have hearing organs.

No measurements are included with the photographs but a general idea of size is given. Tiny moths are those with a wingspan less than 5 mm; very small moths have a wingspan up to 7 mm; small moths are up to 4 cm; medium-sized moths are up to 5 cm; and large are up to 9 cm with very large bigger than 9 cm. Moths of the one species can vary significantly in size. A reasonable guide, with practice, is to compare the size of the scales on the wings with the size of the wings. Moths may be as small as 3 mm wingspan or as large as 23 cm, but the size of the wing scales varies much less. Compare the images of Nepticulidae with those of Saturniidae; the larger the wing scales look compared to the wing then the smaller the moth.

Most larval foodplant records are old and date from when the plant genera *Allocasuarina* and *Corymbia* were not in use, and unless the species was recorded, *Casuarina* and *Allocasuarina*, and *Eucalyptus* and *Corymbia* cannot be distinguished.

The head of *Burgena varia*, (Noctuidae) shows the proboscis, antennae, eyes and labial palpi.

Micropterigidae

- tiny
- hairy head
- short, thickened antennae held up and out
- shining colours
- wings held steeply roof-wise

These moths are the most primitive moths alive today. Amber fossils from the Lower Cretaceous (120 million years ago) found in Lebanon can be placed in this family. They have chewing mouthparts, primitive-type scales, and forewings and hindwings of similar shape and venation.

Micropterigids are very small, shining in gold and blackish-purple and are found in moist places, usually rainforest. There are some Tineidae (pp. 50–51) that look similar but the Micropterigidae can be distinguished by having no spurs on the middle leg. Some have larvae that feed on liverworts and other periphyta but little is known of the Australian species, which may have soil-living larvae.

The moths are active during the day in dappled sunlight and shade where they may be swept from low vegetation with a net. Some, particularly *Sabatinca sterops* from northern Queensland, which is very small and golden in colour, can come to light at night in large numbers. The adults feed on pollen or fern spores and often aggregate at suitable food sources.

In Australia there are about nine species found in rainforests from Tasmania to Cooktown. More than 100 species are known from all continents but they are particularly well represented in New Zealand and New Caledonia.

Sabatinca sterops is found in rainforest from Cooktown to Mission Beach, Queensland. It can be very common but nothing is known of its biology. One of the most primitive of moths, its predominantly golden colour is unusual in this group in Australia. It frequently comes to light at night. Photo: George Gibbs

Sabatinca porphyrodes is found only on the Atherton Tableland, Queensland, where it lives in very wet rainforest. Most of the Australian species of Micropterigidae are purplish-black with cream bars. Photo: George Gibbs

Agathiphagidae

- small
- elongate, hairy head
- antennae held together, directly forward
- wings held steeply roof-wise

These moths have many primitive features like the Micropterigidae: they have chewing mouthparts, primitive scales, and forewings and hindwings of similar shape. The wings are a little longer and narrower than in Micropterigidae and are dark brown in colour with obscure darker markings. At rest they appear very like caddisflies.

One species is found in Australia, where it occurs in the rainforests from Cooloola to Gympie and also in the rainforests between Ingham and Cairns (Queensland). Little is known of adult behaviour but one has been taken at light. Larvae were discovered when wild seed of the kauri pine was collected for forestry purposes. Most of the seed would not germinate and was found to be infested with the larvae, each of which ate out the seed and then remained in a hardened cell where the seed had been. These fell to the ground from the cone as it matured. The larvae proved almost impossible to rear but eventually it was found that the pre-pupal larvae had a resting stage that could continue for as long as 12 years and was very difficult to break. Few kauris are now left in the wild, seed is collected from isolated plantations and the moth appears to be very rare now.

Only two species of this family are known in the world. The other species is found in the Solomon Islands, Vanuatu, Fiji and New Caledonia.

Agathiphaga sp. looks very like a caddisfly but the well-formed scales, rather than hairs, are not caddisfly-like. The moth was photographed on the Atherton Tableland, Queensland, where it had never before been collected.

Hepialidae (Swift Moths, Ghost Moths)

- large
- short antennae
- forewings and hindwings of similar shape
- wings held steeply roof-wise

The adults of most species fly in autumn but a few fly in spring. The flight coincides with autumn rains and frequently hundreds of moths will be seen on a wet night with none or few before that night and none or few after. The adults live for only one day and cannot feed or drink. The base of the hindwing may be suffused in pink, red or mauve but these colours fade rapidly after death.

The females produce vast numbers of eggs, which they scatter on the leaf litter or grass. The larvae live concealed under-ground, either feeding externally on roots (*Abantiades* and *Trictena*), or feeding on leaf litter or the bases of grasses (*Oncopera* and *Oxycanus*). The pupal shells protrude from holes in wood or the soil when the moth emerges.

The larvae of the huge *Zelotypia stacyi* bore in the trunks of eucalypts and those of *Aenetus* bore in various trees, weaving a broad silk covering, incorporating sawdust and frass particles over the entrance to the bore, which is usually (but not always) removed before pupation and replaced by a hard, presumably regurgitated, plug in the throat of the bore.

Hepialid larvae usually bore down-wards. Larvae of *Aenetus* may have an early phase where they feed on fungi in the leaf litter, only later ascending a tree to become a borer.

Australia has a very rich, spectacular and prominent fauna of hepialids with about 120 species of a world fauna of about 500 species.

Aenetus splendens is found in rainforest and wet eucalypt forest from Kingaroy, Queensland to Wollongong, New South Wales. The larvae bore in saplings of many trees and protect the entrance to the bore with a large hemispherical curtain of silk and sawdust. The larvae bore downwards and when they pupate, unlike other species in the group, they do not destroy the silken curtain. This is a male.
Photo John Stockard

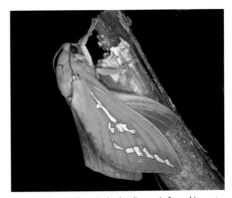

Elhamma australasiae is found commonly from the mountains west of Mackay, Queensland to eastern Victoria. It flies in late summer and early autumn. The larva is thought to feed underground on the roots of grasses and sedges. This is a female.

The very large *Abantiades hyalinatus* is found in wet mountain forests from southern Queensland through to eastern Victoria and Tasmania. It may be grey-brown, dirty yellow, orange to orange-red in colour and the silver streaks may be absent. The mauve colour on the hindwings fades rapidly. The larvae feed underground on the roots of *Eucalyptus*. This is a male.

Aenetus lewinii is found from central Queensland to southern New South Wales. The larva bores in the stem of saplings constructing a web of silk and sawdust around the entrance. This species has been found on *Casuarina* and *Leptospermum*. It comes to light reluctantly and may be more easily reared from stems collected when the larva pupates. This is a female. The male is very different—a light green with some cream lines on the wing.

Aenetus mirabilis is found in the rainforest from Cooktown south to Innisfail, Queensland. The larvae bore in the trunks of rainforest trees. The genus is also found in New Zealand, New Caledonia and New Guinea and contains many spectacular species. Most of the species are found in Australia. This is a male.

Trictena atripalpis is very common through the southern half of Australia extending, along streams, even into arid areas. The adults cannot feed and often fly in large numbers when it rains in autumn or early winter. The larvae feed underground on the roots of eucalypts and are one of the larvae, called bardi grubs, used widely for fishing along the Murray River and its tributaries. This is a male. Photo: Bob Jessop

The very large *Zelotypia stacyi*, with its false eye markings, is said to resemble the head of a goanna. It is found in wet sclerophyll forests from near Brisbane, Queensland to Kiama, New South Wales. The giant larvae bore in the trunks of sapling eucalypts—here the moth is drying its wings at the entrance of the bore but with the empty pupal case removed. This is a female. Photo: Bob Jessop

Fraus pteromela is found widely in southern Australia from Stanthorpe, Queensland to Perth, Western Australia including Tasmania. These moths are a primitive group traditionally placed in the Hepialidae but with features suggesting they do not really belong in this family. This is a male. Photo: Peter Marriott

Nepticulidae

- tiny
- hairy head
- wings folded flat over body
- usually black, or black with a white bar across wing
- antennae held back along leading edge of wing
- base of antenna a broad eye-cap fully covering eye at rest

These minute moths are the most primitive to have forewings and hindwings with markedly different venation, although the venation is greatly reduced because of their small size, and with a frenulum-retinaculum wing coupling system. The related Opostegidae (see p. 44) also has eye-caps but the vast majority of Nepticulidae are predominantly black, while Opostegidae are usually white or silvery.

Nepticulids come to light but they are so small they are difficult to see in the field. Most are reared from larvae forming long sinuous mines ending in a blotch within a leaf. To rear them, the mines must be collected at the right time and kept in a small container.

There are two subfamilies: one, the Pectinivalvinae, is purely Australian with larvae mining in leaves of Myrtaceae; the other subfamily, Nepticulinae, is worldwide but still well represented in Australia. There are only about 20 species named from Australia but the moths are very common and no doubt several hundred species will be found.

The family Nepticulidae contains the smallest moths, sometimes with a wingspan as small as about 3 mm. Most species are dark brown or black, sometimes with a white bar across the wing. The larvae mine in leaves forming a characteristic sinuate track which often widens into a blotch.

Opostegidae

- tiny
- smooth head
- wings folded flat over body
- usually white or silver in colour
- antennae held back along leading edge of wing at rest
- base of antenna a broad eye-cap fully covering eye at rest

These are minute moths, although a few Australian species are 'giants' and are small rather than minute. They occur throughout Australia but are particularly plentiful in some moist tropical savannah habitats.

The Opostegidae have eye-caps like Nepticulidae, forewings and hindwings with different but much reduced venation, and a frenulum. Their biology is unknown in Australia but a few have been reared overseas from very long tortuous mines in sapwood. Only 18 species are listed from Australia, but there are probably hundreds.

The Opostegidae have not been studied in Australia but come commonly to lights. They are so small they are difficult to collect and preserve and, unlike Nepticulidae, they have not been reared.

Most opostegids are silvery with little streaks or a spot near the wing tip. There are many undescribed species in Australia and some species in the north can be very common.

Heliozelidae

- tiny
- smooth head
- antennae held back along leading edge of wing at rest
- black with white or silver bar
- wings held roof-wise

These are tiny moths but a few southern Western Australian species are larger and golden in colour. They are without eye-caps and have smooth, scaled heads that are rather dome-shaped but narrow towards the mouthparts and the scales run onto the base of the short proboscis. The venation of forewings and hindwings is different and reduced.

They are often active during the day, sometimes visiting flowers, but some come to light at night. The larvae are miners in leaves and when fully developed form a case from the upper and lower surfaces of the mined leaf strongly silked together. This they cut out of the leaf and fall to the ground. When there are a lot of them they leave the leaf with a shot-holed appearance.

Heliozelids are found throughout most of Australia. There are about 35 recognised species and over 100 species worldwide, but many undiscovered ones exist.

This tiny, unidentified, female moth came from the Hartz Mountains in Tasmania. Nothing else is known about it. Photo: Andreas Zwick

Adelidae (Fairy Moths)

- small
- smooth head
- extremely long hair-like antennae
- antennae held forward and outward at rest
- wings held roof-wise over body
- some species are grey but others are shining metallic golden or green

One group of fairy moths has wings with bright metallic scales in green, yellow and black and is active in the day, usually seen feeding at flowers. These also have distinctive, large eyes set close together. Another group is grey and flies at night.

The larva initially feeds within a developing flower and later constructs a portable case of leaf or flower fragments and drops to the leaf litter where it continues to feed on fallen flowers.

In Australia there are three species of the night-flying subfamily, Nematopogoninae, all in the south. The metallic Adelinae, with eight species, are widespread with most in northern Australia. One species, *Nemophora chrysolamprella*, is widespread and the adults are often seen on the flowers of *Bursaria spinosa* (Pittosporaceae) upon which the larvae feed. Others feed on wattles.

There are about 300 species worldwide but they have not been studied in Australia.

This small male fairy moth is a species of *Nemophora*. It is found from Heathlands to Iron Range on Cape York Peninsula, Queensland. These glittering little moths are day-flying and visit flowers but also occasionally come to light at night. Nothing is known of its early stages.

Palaephatidae

- small
- woolly head
- wings held steeply roof-wise
- antennae held back along leading edge of wing

The Palaephatidae is the most advanced family to retain the system of a single female opening. They look very like tineids and cannot be distinguished without dissection. They fly at night.

There are two very different genera in Australia. *Azaleodes*, with four species, has broad wings, is golden brown with fine black spots in the male, and is found only in rainforest from Wollongong, New South Wales to Cooktown, Queensland. The other genus, *Ptyssoptera*, is more widely distributed and has narrower wings.

The larvae of *Ptyssoptera* are leaf miners at first but later form shelters by silking together leaves of the foodplant. The few that have been reared fed on Proteaceae.

The family is found only in South America and Australia. Altogether there are about 60 species with 11 described species in Australia.

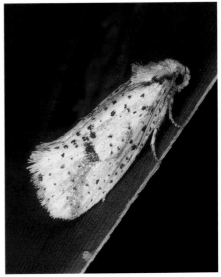

Azaleodes fuscipes is found in Queensland rainforest from Cooktown southwards, nearly to Townsville. Nothing is known of its biology. This is a male.

Psychidae (Case Moths, Bag Moths)

- small to medium size
- small, hairy head
- wings held steeply roof-wise
- antennae thread-like or broadly pectinate with pectinations shortening well before tip
- antennae held back along leading edge of wings
- usually with a very long abdomen, much longer than wings
- the medium-sized species have wings sparsely scaled or unscaled

Female psychids have an abdomen with two openings; this is the most primitive family with this feature. Psychids also lack a proboscis and cannot feed.

The family is characterised by the elaborate larval cases or bags, often with neatly trimmed sticks or grass stems placed in parallel, or sometimes with untidy bits of leaf or twig attached at random. Each species has its own characteristic case. The larvae feed on a wide range of plants and lichens. They are very rapid flyers, some in the daytime, most at night.

There are two very different groups of psychids. One group has not been studied in Australia but overseas it contains several subfamilies. These usually have longer wings than those in the subfamily Psychinae, and the wings have colour patterns usually in grey but sometimes in black and white or gold and in one species, *Cebysa leucotelus*, the female is metallic blue and gold. In this group the females are fully winged and the antennae may be simple or pectinate in one or both sexes.

In strong contrast, the Psychinae have females that are wingless and may never emerge properly from the pupal shell that remains within the larval case. Fertilisation is effected by the very flexible and extensible abdomen of the male penetrating the case

Hyalarcta nigrescens is a small moth, occurring widely from northern Queensland to southern New South Wales. This is one of a group of case moths in which the female is wingless and does not emerge from the larval case. The larvae disperse and feed on a very wide range of plants. This is a male.

and the pupal shell of the female. The female is detected by the chemicals (pheromones) she produces to attract the male and the male has broadly pectinate antennae to detect the chemicals.

Psychinae have a stout thorax and short triangular wings that have little or no colour pattern. They may be white, grey or golden brown. With immobile females, dispersal is carried out by the freshly hatched larvae that may spin silk and float off in the wind.

The family is found worldwide with about 1000 species. In Australia there are about 180 species, which are largely unstudied.

This medium-sized *Lepidoscia* species comes from southern New South Wales. There are large numbers of unstudied and unidentified *Lepidoscia* in Australia. Their larvae build conspicuous cases of silk and plant debris, which they carry around. Most species feed on lichens but others feed on a variety of plants.

This case is characteristic of the genus *Lepidoscia*. Case moths are frustrating to rear because of very high rates of parasitism.

This small *Iphierga* species is from northern Queensland but a number of very similar-looking ones are found widely in Australia. The caterpillars live in cases made of silk and pieces of plant debris, which they carry around. This is a male.

This case is made of silk and is decorated with large sections of cut leaf. Not only does the larva drag its case around with it, but it must also increase the size of the case as it grows and continue to decorate it with leaves.

This large, bright, psychid moth, *Lomera pantosemna*, is found in the moist southern forests of Western Australia from Walpole to the Fitzgerald River. This is a male. The female is as yet unknown but will be wingless. Nothing is known of the early stages.

Tineidae (Clothes or Wool Moths)

- small to medium size
- woolly heads
- wings held roof-wise
- thread-like antennae held back along leading edge of wing at rest
- often bristle-like scales protruding from palps
- moths run rapidly on alighting

There are 11 subfamilies of Tineidae in Australia, two of which have smooth-scaled heads and are usually very small moths. The adults are nocturnal but many rarely leave the hidden refuges where they breed. The larvae may make cases out of detritus and frass silked together or may live in silken tunnels in the substrate.

The subfamily Tineinae contains the well-known clothes moths with about 10 species that feed on wool, feathers or fur. Wool is a marginal food, which usually has to be dampened to allow some fungal growth, or enriched in some other way before it is attacked.

Some species are detritus feeders, some feed in rotting wood, one in bat guano in caves, and others feed in fungi. All these seem to have fungi as a common component and it may be that tineids are basically fungal feeders with some specialising in dead plant or animal materials.

One or two species of *Opogona* damage live buds in situations where they are close to plant detritus. Some tineids have been recorded from wine corks.

Worldwide there are about 3500 species. In Australia there are about 189 described species with several hundred more known to be present.

The small *Micrerethista entripta* is found from north of Cooktown, Queensland to Wollongong, New South Wales. Nothing is known of its biology but the larva of a related species in South-East Asia has been found tunnelling in a log. The group to which it belongs is unique in the tineid family in having hearing organs.

This small *Edosa* species is one of a large group of similar species that are common throughout Australia. They all have a similar pattern of yellow and brown or black. It is remarkable for such a plentiful group that none of the larvae have ever been found.

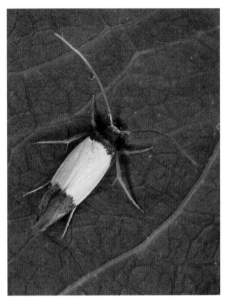

The moths of the genus *Erechthias* are very numerous in Australia and the islands of the Pacific. Some of them have a small or large (as in this case) part of the wing tip turned up. The larvae of many other species of this genus feed on dead plant detritus but the unidentified species shown here has never been reared. It is found in the rainforests of north-eastern Queensland.

This small *Opogona* species is commonly found from north of Cooktown to Cardwell, Queensland. Nothing is known of its biology. Other species in this genus feed on dead plant debris. This is a male.

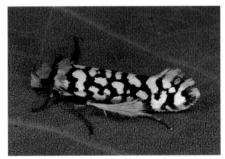

Gerontha acrosthenia is striking because of its very long hind legs that are characteristic of the genus. It is found in rainforest in New Guinea and from Cape York to just north of Townsville, Queensland. There are also records from near Brisbane. The antennae of this specimen have been broken off short when it was caught. One species overseas has been reared from larvae boring in dead wood.

The large, boldly patterned *Moerarchis clathrata* is found in southern Western Australia from Geraldton to Thomas River. It is closely related to *M. australasiella* from eastern Australia and also has larvae that bore in the dead trunks of grass trees (*Xanthorrhoea*, Xanthorrhoeaceae).

Clothes moths and meal moths

The clothes moths belong in the family Tineidae and are all small moths that shun light and have retiring habits. Most people never see them and they are only detected when their damage is discovered. In Australia there are seven species of *Tinea*, one *Tineola* and one *Trichophaga* that are widespread overseas and have entered Australia with humans. As well, a naturally occurring species, *Monopis crocicapitella*, may cause damage when it strays from nests, animal bodies and bat guano upon which it usually feeds.

Clothes moths all attack wool, any woollen or hairy fabric, furs, carpets, feathers, down, and clothes. All the introduced species are light brown or golden in colour except for *Trichophaga tapetzella*, the tapestry moth, which is black and white and named for its colour pattern rather than any preference for tapestries.

Clothes moths thrive in dark, undisturbed parts of the house. The larvae (the destructive

The clothes moth, *Tinea dubiella*, (Tineidae), can cause immense damage when it attacks stored clothes in dark, undisturbed corners particularly if the wool has been enriched by a little mould or nutrients at some earlier stage. Usually the cases and damage are the first clue to trouble as the moths are rarely seen. Photo: John Green

stage, the adults do not feed) can digest keratin but, as wool or hair is a very poor source of nutrients, the moths do better if the wool or hair has been enriched in some way. Carpet edges may occasionally become moist near the walls, or a sweaty jumper may be put away unwashed. Such items behind a sofa or in a dark undisturbed cupboard are favourite places for attack.

Clothes moths are best discouraged by spring-cleaning, not putting away dirty items and moving the furniture regularly. The best control is putting small, infested items in the freezer for a week and cleaning up larger things, exposing them to sunlight, air and activity.

Meal moths are much more common than clothes moths and most pantries are infested from time to time. These are moths in the family Pyralidae and are small, but larger than clothes moths, and are mostly grey or buff in colour. They usually enter the pantry as eggs, too small to see, brought in on packets of food. They feed on a wide range of products including cereal, grain, meal, wheat germ, bran, flour, nuts, biscuits and chocolate. The best way to control an infection is to clean out the pantry, disposing of infected food and freezing everything that may be infested. To prevent further infestation susceptible foods should be stored in sealed containers such as screw-topped jars. If a jar is infested because eggs were introduced with the contents then the infestation cannot spread to other jars. Remember that infested food is a sign of healthy food; the moths would not be there if it had been chemically contaminated. Contamination should always be considered before any chemical control is contemplated.

These are the larval cases of *Tinea dubiella*. There are seven pest species of *Tinea* in Australia, all with larvae that live in cases made of silk, wool fragments and droppings. All the clothes moths are cosmopolitan and have been introduced to Australia. Photo: John Green

This is *Pyralis farinalis*, family Pyralidae, one of the meal moths frequently found in pantries. It is cosmopolitan and was probably brought to Australia early in settlement. Most other meal moths are dull grey in colour. Photo: Ian Common

This clothes moth, *Tineola bisselliella*, (Tineidae), is very small and rarely seen because of its secretive habits but can be found throughout Australia. Most people mistake other harmless moths for clothes moths. Almost all are tiny, brown or golden. The different clothes moths are very difficult to tell apart and need expert identification. Photo: Ros Schumacher

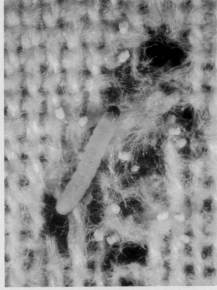

This species of clothes moth, *Tineola bisselliella* (Tineidae), has no larval case and can be seen here feeding on woollen fabric. This is one of the commonest moths causing damage in Australia. Photo: Ros Schumacher

Galacticidae

- small
- smooth head
- short, broad, dumpy bodies
- wings held low-pitched roof-wise
- antennae short, usually thread-like, held back against leading edge of wing
- grey or white, speckled grey

Most galacticid species in Australia are found in arid central Australia and in Western Australia. The adults come to light.

This small family has four species of *Homadaula* in Australia but little is known of them. They have been reared from *Acacia* shrubs upon which the larvae form a massive silken web enclosing many phyllodes. The fully fed larvae leave the web to form solitary cocoons on the ground.

Thirteen species occur in Europe and Asia.

Homadaula myriospila is found from Shark Bay to Cape Arid National Park, Western Australia. The larvae web the phyllodes of *Acacia* together with silk and feed on the leaves in this shelter. Most of the species are found in the arid areas of central and north-western Australia. Photo: Ted Edwards

Roeslerstammiidae

- small
- smooth head
- wings held steeply roof-wise or wrapped about body
- antennae long and simple, held out and forward from head
- palpi upcurved

This small family, with a total of 40 species, is found only in Europe, Asia and Australia. In Australia there are 23 recorded species.

In one rainforest genus, reared from *Elaeocarpus reticulata* (Elaeocarpaceae), the eyes are divided horizontally by a band of scales, a feature not found elsewhere in the Lepidoptera.

The moths in this family mostly come to light. The larvae feed on green leaves and, at least in later instars, silk leaves together to form a shelter. They feed on Epacridaceae and others on Proteaceae and one day-active, bronze metallic-coloured species in Tasmania feeds on *Nothofagus*.

Chalcoteuches phlogera is found in the alpine areas of Tasmania where the larvae feed on the Tasmanian myrtle, *Nothofagus cunninghamii* (Fagaceae). The moths fly in the sunshine. This is a female. Photo: Andreas Zwick

Bucculatricidae

- tiny
- woolly heads
- wings held steeply roof-wise
- antennae held back along leading edge of wings
- hindwing very narrow with long hair scales from trailing edge
- small, exposed, ridged cocoons

These moths are so small that most are unlikely to be noticed and they are usually reared. The larvae start life as leaf miners but later live under the leaf and erode irregular sections of the leaf tissue. When the larva sheds its skin it constructs a special moulting cocoon. The final cocoon is most distinctive, being elongate, cylindrical and with a series of longitudinal ribs and furrows. It is placed on the underside of the leaves or, in *Ogmograptis*, in the leaf litter.

The larvae feed on a variety of plants, but one common group can occur in large numbers on *Commersonia* (Sterculiaceae) and *Pomaderris* (Rhamnaceae). The most conspicuous are the members of *Ogmograptis*, which cause the well-known scribbles in the bark of many smooth-barked eucalypts, which gives *Eucalyptus haemostoma* the common name of 'scribbly gum'. While many species of eucalypts have scribbles, very little indeed is known of the moths, the numbers of species or their biology (see p. 57).

The bucculatricids are unstudied and only 13 species have been described from Australia but there must be many more. There are about 250 species worldwide.

The grooved cocoons with an extruded pupal shell are typical of this family. Here a species of *Bucculatrix* has spun the cocoons on the underside of a leaf of *Commersonia* (Sterculiaceae).

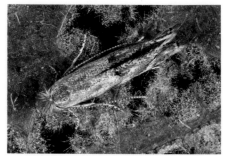

Here the same *Bucculatrix* species rests on the leaf of the foodplant damaged by the larvae. Its colours closely match those of the leaf.
Photo: Natalie Barnett

Scribbly Gum Moths

Many smooth-barked eucalypts are well known for the scribbles in the surface of the bark, but it is less well known that they are caused by the larvae of moths. Scribbly gum moths appear to fly for a very short period, are reluctant to come to light and are very rarely collected. They are also extremely difficult to rear as larvae, and we know almost nothing about them.

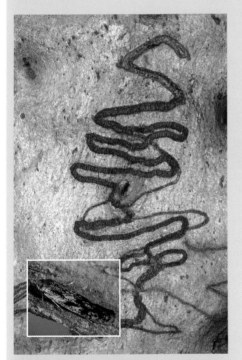

The scribbles on many eucalypts are caused by the larvae of species of *Ogmograptis* moths, (Bucculatricidae). The larvae mine in the bark, producing a zigzag gallery, then emerge to form cocoons in litter near the tree.

Inset: This scribbly gum moth, *Ogmograptis,* has been reared from scribbles on *Eucalyptus rossii*.

Photo: Natalie Barnett

It is thought that the moth lays the egg onto the bark. The larva commences a sinuous, zigzag mine that widens gradually as the larva grows and when somewhere about half grown it suddenly reverses course and mines back parallel to the old mine. Why it does this is unknown but there are several possible reasons. It may escape predation or parasitisation if the parasite or predator targets the end of the mine. There may also be a nutritional benefit from mining near the old scar tissue.

When fully grown, the larva emerges from the trunk of the tree and forms a grey, characteristically ridged bucculatricid cocoon under bark at the base of the tree or in the leaf litter. The length of the life cycle is unknown but one year is likely.

The small grey bucculatricid, *Ogmograptis scribula*, was reared from cocoons associated with scribbles on *Eucalyptus pauciflora* (snow gum). The moths from this eucalypt emerge in February or March but those from other trees may do so at other times.

As the tree grows, the outer bark is sloughed off each year exposing the old mines.

Scribbly gum moths are difficult to rear and very little is known of their biology. Most have narrow wings, long hair scales on the trailing edges of the wings and sombre colours. This has a wingspan of about 8 mm. Photo: John Green

Gracillariidae

- small
- usually with smooth head but some hairy
- wings held back along body steeply roof-wise
- antennae often long, held back along wings at rest
- both forewings and hindwings very narrow with long scale fringes from trailing edge

These are small, slender, elegant moths, often very brightly coloured and intricately patterned. Moths with parallel-sided forewings, about five times as long as they are broad, are probably gracillariids, but not all gracillariids have wings as long as this. The labial palpi are usually curved upwards but sometimes droop. The legs are very long and slender.

The larvae are leaf miners in a very wide variety of plants starting as sap-feeders, puncturing cells and ingesting the contents, but later they feed on the leaf tissue forming a tortuous and then blotch-shaped mine. A few small species remain as sap-feeders throughout life and these have a non-feeding final instar larva that is specialised to form the cocoon. Larvae normally pupate in the leaf mine. The larvae of one southern Australian group of giant gracillariids form terminal stem galls in *Eremophila* (Myoporaceae) (see p. 32).

Gracillariids are mostly too small to observe in the wild but some come to light and have a variety of resting positions. In one group the head is held very high, legs are splayed and the body, enclosed in the wings, slopes down to the substrate at the wing tips. Adults are usually reared from mined leaves. There are about 190 described species but many remain undiscovered. The family is worldwide in distribution with about 2000 species.

This very small, elegant moth, photographed in southern Western Australia, is probably a species of *Caloptilia*. They are very poorly known in Australia. The larvae of this genus probably mine in leaves.

Macarostola formosa is a minute but brilliant moth, which is found from Brisbane, Queensland to Sydney, New South Wales. The larvae mine in the leaves of the tree *Syzygium francisii* (Myrtaceae). Photo: John Stockard

Yponomeutidae

- small to medium
- smooth or, more rarely, woolly head
- wings held roof-wise or wrapped around body at rest
- long simple antennae held back along leading edge of wing at rest

This family contains a diverse array of moths, some of which may eventually be moved to other families. The palpi may be short or long and often somewhat upturned. Usually the wings are long but fairly broad.

The larvae of some species live in a communal silken web, spun in the leaves of the foodplant, where they also pupate. In others the larvae live singly, webbing over leaves or flower buds upon which they feed. The genera *Yponomeuta* and *Atteva* are medium-sized moths; the former has forewings of pure white with an ermine-like speckling of black spots, while the latter has orange brown forewings with numerous white spots. Others may be quite small and grey.

There are about 60 species in Australia and the family has about 600 species worldwide.

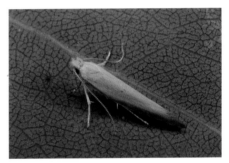

The small *Zelleria isopyrrha* is found from Busselton to Albany, Western Australia. Nothing is known of its biology. This is a male.

The medium-sized moths in the *Atteva* genus are contrastingly spotted with white. This moth is found in rainforest and monsoon forest in the top end of the Northern Territory and in Queensland, south to Lismore, New South Wales. The larvae of one species spin a communal web in the leaves of *Polyscias murrayi* (Araliaceae).

Yponomeuta paurodes is found from the Torres Strait Islands, and in Queensland, south to Grafton, New South Wales. These small moths are called ermine moths overseas. Their larvae have been found singly on small webs in the leaves of *Cassine australis* (Celastraceae).

Argyresthiidae

- very small
- head with roughened scales
- wings held roof-wise back over body at rest
- simple antennae held out from head
- both wings very narrow, with long scale fringes from trailing edges

These moths sit with the head down near the substrate, with their body held up at a steep angle. Overseas the larvae are known to bore in the buds, stems and fruit of trees.

There are about 150 species worldwide but only one species in Australia.

Argyresthia notoleuca is the only species of the family in Australia. Nothing is known of its biology. This very small moth is found in rainforest in the Cairns district and the Atherton Tableland south to near Townsville.

Plutellidae

- small
- head smooth or rough-scaled
- wings usually held roof-wise over body
- antennae usually held together extending directly in front of head

This family includes some genera that may belong elsewhere but are placed here until work is done on them. The adults usually come to light. The larvae of some feed on leaves beneath a slight web of silk. Others are known to feed on the stems and bulbs of orchids.

The very common cabbage moth (not to be confused with the large black and white cabbage butterfly) has an enormous distribution all around the world. About 25 species are known from Australia and about 50 species worldwide.

Plutella xylostella is also called the cabbage moth or diamond-backed moth, and is a significant pest of crops in the cabbage family. It also feeds on natives and weedy species of the same plant family and is found throughout Australia. One of the world's most wide-ranging moths, it is abundant on all continents and even isolated islands.

Tritymba pamphaea has characteristic, forward-pointing antennae. It is found from southern Queensland to western Victoria. Its biology is unknown but several related species have been reared from tough cocoons found under the bark of eucalypts.

Glyphipterigidae

- small
- smooth head
- wings held roof-wise
- simple antennae held back along leading edge of wings at rest
- labial palps gently curved upwards
- wings fairly broad

Glyphips are characterised by shining grey or black wings, sometimes with white or yellow patches, and with a pattern of shining metallic blue or green scales, particularly towards the tips of the wings, where there may be an intricate swirl of metallic scales.

The adults are active by day and rarely come to light. They are usually swept from the vegetation with a net or found on flowers. The larvae of one species are known in Australia but some other species have been reared from sedges and adults are often found clearly associated with grasses or sedges.

One giant species with a wingspan of 60 mm has recently been described from Tasmania, where the larvae bore in the shoots of the giant pandani, *Richea pandanifolia* (Epacridaceae).

All glyphips overseas feed on grasses or sedges, boring in buds, stems or seed heads. In the Snowy Mountains one species is common in autumn on *Carex gaudichaudiana*, and *Glyphipterix chrysoplanetis* is often associated with the grass *Microlaena stipoides*. There are 58 species in Australia, mostly from the cool and moist southern parts, although one is associated with cane grass around Lake Eyre. There are about 380 species worldwide.

Glyphipterix meteora is found from Brisbane, Queensland to Adelaide, South Australia, including Tasmania. Nothing is known of its biology. Moths of this family usually feed, as larvae, in grasses or sedges. They are day flying with brilliant reflective scales that shine in the sun. Photo: Axel Kallies and Peter Marriott

A male and female *Glyphipterix chrysoplanetis*. Glyphips often feed at daisies or other small flowers in the daytime. This species is found from just north of Brisbane, Queensland to Tasmania. Most species are found in the moist south-eastern part of Australia. Photo: Axel Kallies

THE GELECHIOID FAMILIES

The gelechioid group, encompassing all the families from Oecophoridae to Scythrididae (see pp. 64–90), is distinguished by having sharp-pointed, upturned and sickle-shaped palpi (see the illustration on page 36). Some other families have similar upturned palpi but usually they are less sharp-pointed and a very few gelechioids have different palpi. The base of the proboscis has scales on it, which are visible only under a microscope. The pupal shell is not extruded from the cocoon when the moth emerges. The gelechioid families are very difficult to distinguish and the family classification is very unstable and controversial. There are three species-rich families in Australia: the Oecophoridae, Cosmopterigidae and Gelechiidae. A large proportion of the small moths seen at light belong to these families.

Oecophoridae

- small, occasionally medium-sized
- usually smooth head
- wings held roof-wise or flat back over body
- simple antennae held back along leading edge of wing at rest
- upturned, sharp-pointed, sickle-shaped palpi
- hindwing usually lanceolate
- adult has a waddling gait

This hugely diverse group in Australia has three subfamilies: Oecophorinae, Stathmopodinae and Stenomatinae. The Oecophorinae have lanceolate hindwings with short scale fringes. The Stathmopodinae usually have narrow hindwings and long scale fringes on the trailing edge, and they often rest with the hind legs splayed out almost at right angles to the body forming a cross. The hind legs have whorls of long spine-like scales at the joints (p. 68). The Stenomatinae are grey in colour and rest with their wings flattened roof-wise. Their hindwings are much broader than the forewings; almost a full quarter circle in shape.

The larvae of Oecophorinae typically make shelters by tying leaves together with silk or making silken tunnels amongst leaves, or cutting cases from leaves, but their habits are diverse. Many Australian species feed on the dry myrtaceous leaf litter on the forest floor; others feed in animal scats or nests. The larvae of the Stenomatinae make shelters tying leaves together.

Some Stathmopodinae feed in rust galls or spore-bearing bodies of ferns and one is carnivorous and feeds on scale insects. The adults often mimic wasps or poisonous beetles.

The Oecophorinae have explosively radiated in Australia with over 2300 named species but probably more than 5000 altogether. The rest of the world has fewer than 1000 species.

This species of *Habroscopa* is found in rainforest from the Bloomfield River, south to the Cairns district, Queensland. Its biology is unknown. These and some related genera have heavily scaled forelegs, which are displayed when at rest. Subfamily Oecophorinae.

This species of *Agriophara* is found near Melbourne, Victoria. Nothing is known of its biology but related species have been reared from leaves silked together on eucalypts. Subfamily Stenomatinae.

Photo: Peter Marriott

The brilliant *Habroscopa iriodes* is found in rainforest in New Guinea and in Australia, from Cooktown, Queensland to Hobart, Tasmania. Nothing is known of the early stages. This is a male. Subfamily Oecophorinae.

Scatochresis innumera has a most distinctive, finely lithographed pattern. It is found from the Atherton Tableland south to Brisbane, Queensland in eucalypt forests. The moth has been reared from a larva feeding in the scat of a possum. This is a male. Subfamily Oecophorinae.

Stathmopoda melanochra is common from southern Queensland to Tasmania and South Australia. It has very narrow wings. Unusually for a moth, the larvae are carnivorous and feed on scale insects, forming silk tunnels amongst the shells of their consumed prey. Subfamily Stathmopodinae.

This species of *Chrysonoma* is found in central and southern New South Wales in sclerophyll eucalypt forests. Like some related species, the larva may feed between two silked-together living eucalypt leaves. Subfamily Oecophorinae.

Piloprepes antidoxa is found in eucalypt forests from southern Queensland to south-western Western Australia, and even into semi-arid parts like Broken Hill, New South Wales. The conspicuous colour pattern, believed to mimic bird droppings, occurs frequently in moths. Subfamily Oecophorinae.

Hemibela callista is found in eucalypt forests and woodlands from southern Queensland through New South Wales to southern Western Australia. The larva feeds on eucalypt leaves and constructs a shelter by neatly severing a dead twig about 2 cm long and boring through the core for the full length of the twig. It lives in this shelter that it carries perpendicular to the leaf and remains inside, with its head out while feeding. Subfamily Oecophorinae.

Agriophara cinerosa is a medium-sized moth found in moist eucalypt forests in southern New South Wales and Victoria. When resting on the fibrous bark of trees, its wing pattern conceals it effectively. The larvae have been reared from shelters of eucalypt leaves made by tying two green leaves with silk. This is a male. Subfamily Oecophorinae.

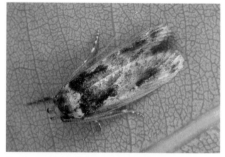

Barea consignatella is found from southern Queensland through New South Wales and the Australian Capital Territory, usually in moist eucalypt forest. The larvae live in silken galleries and feed on moist rotting dead wood, particularly of eucalypts. This moth is worn and part of the pattern on the wings is lost. Subfamily Oecophorinae.

Garrha zonospila is found in eucalypt forests from Lismore, New South Wales to Gisborne, Victoria. Nothing is known of its biology but the larvae of other species in the genus feed on dead eucalypt leaves, making a case from two oval cut pieces of leaf. Its finely annulated antennae are a feature of most species of *Garrha*. Subfamily Oecophorinae.

This species of *Aristeis* is found from southern Queensland to southern New South Wales. It inhabits eucalypt forests and is one of the very few Australian oecophorids with close relatives in the Oriental region. The larva feeds on various eucalypts. It makes a spiral case of silk with fragments of leaf and drop-pings that it drags along the leaf while it feeds. This is a female. Subfamily Oecophorinae.

This species of *Psaroxantha*, found on the Atherton Tableland, Queensland, belongs to a large group of small yellow and pink moths, all with a similar basic pattern. Nothing is known of its biology, but larvae of a related species have been reared from lichens. Subfamily Oecophorinae.

This often-collected moth is reminiscent of *Pseudaegeria phlogina* (in the same subfamily) but rather different in details. It is found in the Kimberleys of Western Australia, the top end of the Northern Territory and in Queensland from Coen to Sarina. Nothing is known of its biology. Subfamily Stathmopodinae.

This species of *Zatrichodes* is found in the monsoon forests and rainforests of Queensland from Cooktown to Tully. The halo of 'hairs' are masses of bristle-like scales radiating from the joints in the legs, giving the moth a very peculiar and distinctive appearance. Nothing is known of its biology. Subfamily Stathmopodinae.

This species of *Snellenia* is found in the rainforests of the Atherton Tableland, Queensland. The genus mostly mimics distasteful beetles and this species may also. Nothing is known of its biology. Subfamily Stathmopodinae.

Ancistroneura ammophara is such a poorly known moth that only one has been collected, other than this one photographed. Both were taken near Kuranda, Queensland. This one was found in tropical eucalypt woodland. Subfamily Oecophorinae.

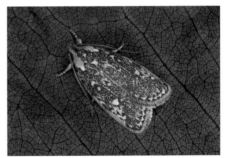

The bold colour and bent wingtips make *Wingia aurata* easily recognisable. It is found throughout the continent, even in central Australia, wherever eucalypts are found. The larva ties two eucalypt leaves together with silk, forms a silken tube between the leaves and feeds on the tied leaves. Subfamily Oecophorinae.

This species of *Euchaetis* is found in Western Australia from the Pilbara south to Albany. Nothing is known of its biology but related species have larvae that tie the green leaves of eucalypts together with silk and feed on the leaves. Subfamily Oecophorinae.

Philobota xanthastis is found in the southern part of Western Australia, from Jurien Bay south and along the coast as far east as Mt Ragged. It flies in spring and summer but nothing is known of the early stages or foodplant. Subfamily Oecophorinae.

This brightly coloured *Crepidosceles chryserythra* is found from the Atherton Tableland in northern Queensland to southern New South Wales. Nothing is known of its biology. Subfamily Oecophorinae.

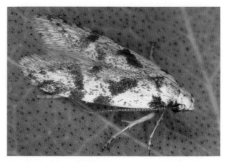

Catacometes hemiscia is found throughout southern Australia north to the Pilbara, Western Australia and to Roma, Queensland. It has not been found in Tasmania. Larvae have been reared from green eucalypt leaves tied together with silk. Subfamily Oecophorinae.

Oxythecta hieroglyphica is common in eucalypt forests from southern Queensland and New South Wales. The moths have been reared in large numbers from the droppings of wallabies. Other moths in the genus have been reared from the droppings of wombats. Subfamily Oecophorinae.

Conobrosis acervata is found in southern Western Australia from Eneabba to the Thomas River. The moths have been reared from larvae feeding on the dead, dried male cones of cycads. Subfamily Oecophorinae.

Piloprepes aemulella is found throughout northern and central Australia and south to Victoria. The larvae tie the leaves of eucalypts with silk, making a shelter from which they feed. Subfamily Oecophorinae.

Scat Moths

In 1992 the threat of forestry operations to a koala population in the Tantawangalo State Forest in southern New South Wales led to an attempt by conservationists to define the range of the koala colony. One possible method was to survey the area for koala scats on the ground but how long did koala scats persist in the wild? This led to the discovery of moth larvae in the koala scats.

More than 30 species in the genus *Telanepsia* feed in the scats of leaf-feeding koalas and possums. Almost all are small- to medium-sized grey moths. About five species of *Oxythecta* have been found in the scats of the grass-feeding wombat, rock wallaby and wallaby. These moths are small and white with an irregular pattern of rich brown. One species of *Ioptera* also feeds on these scats—it is larger and grey with faint longitudinal lines. Undoubtedly many more species remain to be discovered.

The biologies of the various species of scat moths differ greatly, as does their distribution. In more arid areas, the larvae may live in a tunnel in the soil, surfacing at night to feed on the scat. Most, however, live within the scat, hollowing it out completely and eventually pupating within the scat; those in koala scats require only one scat to complete development.

Adult moths seem to prefer to rest on scats suitable to lay their eggs on. The koala scat can be seen as a concentrated, prepackaged parcel of dead leaf and it is not surprising that some of the leaf-litter feeding oecophorids have specialised in this food source. Equally, there are many grass-feeding oecophorids that have possibly moved from grass to the highly concentrated wallaby or wombat scat. Another small oecophorid, *Boroscena phaulopis*, feeds in the nests of finches but nothing more is known of its biology.

Much of what we know about the considerable array of moths that feed in the scats of marsupials has been due to the work of Dr Ian Common.

Telanepsia stockeri (Oecophoridae) is resting possessively on a koala scat. Its larva feeds within a single scat, and when fully grown forms a cocoon within the eaten-out scat. This species is known only from adults reared from koala scats from the Tantawangalo State Forest in southern New South Wales. Photo: John Green

The larva of *Oxythecta acceptella* (Oecophoridae), feeds on the scats of wallabies and wombats as do other species in the genus. This moth is found from Nambour, Queensland to Melbourne, Victoria. Photo: Ian Common

The Golden-shouldered Parrot Moth

The golden-shouldered parrot nests on Cape York Peninsula in the tall, conical mounds of the termite *Amitermes scopulus*. During the wet season, the bird excavates a large nest hollow in the mound with a small entrance for the bird. Egg laying takes place from March to May.

In 1922 William McLennan found that the parrot's nests were infested with the larvae of the small grey oecophorid, *Trisyntopa scatophaga*, which ate the excreta of the nestlings. McLennan observed copulating moths in a mound with newly laid eggs, suggesting that the moths are attracted to newly excavated hollows with eggs. The larvae

live in a mass of silken tunnels on the floor of the nest and consume the nestlings' excreta as they are produced.

When fully grown, the moth larvae form a cluster of cocoons placed horizontally at the thinnest part of the nest wall, so that they extend through the wall from the interior of the nest to the exterior of the mound. This allows the moths to escape even if the termites close the nest entrance after the birds have left.

It would appear that the presence of moth larvae contributes greatly to nest hygiene. While most nests are occupied by moth larvae, some are not, and nestlings are successfully

This termite mound has an entrance hole leading to the nesting cavity of the golden-shouldered parrot.
Photo: Don Sands

The cocoons of *Trisyntopa scatophaga* in the wall of the termite mound extend from the nesting cavity to the outside of the mound. This allows the moths to emerge easily. Photo: Don Sands

The adult *Trisyntopa scatophaga* (Oecophoridae) is a singularly drab moth. Photo: Ted Edwards

reared from moth-less nests. Dr Stephen Garnett has found that the moths are not necessary to the birds, and in one case cocoon masses spun before the nestlings have left the nest have blocked the parrot's entrance and the nestlings have died. However, it is still not clear whether moth infestations on average increase or decrease the breeding success of the parrot.

The moth appears to be strictly associated with the golden-shouldered parrot and has not been found with other birds in the same area nor to live independently of birds on some other food. The moth has never been found at light or anywhere else but in a parrot's nest.

A close relative of the golden-shouldered parrot, the hooded parrot, is found in the Northern Territory and is also associated with a moth that may be the same species, although this is not known for sure. The extinct paradise parrot was closely related to these birds, but we will never know if it was likewise associated with a moth.

A related moth, *Trisyntopa euryspoda*, which is very widespread in Australia, has been reared from the nests of eastern rosellas

and mulga parrots and probably occurs in many other parrot nests but it seems to be a much less regular nest inhabitant.

The golden-shouldered parrot. Photo: Stephen Garnett

Leaf-litter moths

In Australia there are numerous moths in the families Oecophoridae, Lecithoceridae, Blastobasidae, Tineidae, Tortricidae, Pyralidae and Noctuidae whose larvae feed on leaf litter. Most notable of these is the Oecophoridae. This family has radiated explosively in Australia, with the number of species vastly exceeding the number known from the rest of the world.

Many oecophorids have specialised in feeding on the dry-fruited Myrtaceae, such as eucalypts, leptospermums and melaleucas. Some feed on the green leaves of these plants,

Many invertebrates help break down dead leaves. Here a fallen branch has had all the leaves skeletonised and the plant material rapidly recycled. Oecophorids play a vital role in this process.
Photo: Ted Edwards

while others have adapted to feeding on the dry litter of dead leaves on the ground, not only on myrtaceous leaves but those of wattles and banksias as well.

Most of these moth larvae form silken shelters in the dead leaves, eroding the surface of a leaf. A fallen eucalypt branch in south-eastern Australia will often be found where the leaves hanging from the branch are completely skeletonised due to the action of moth larvae. Importantly, the moths attack the hard intact leaves at, or just beneath, the surface of the litter that very few other things will eat. A density of Lepidoptera larvae of 438 per square metre has been recorded in the litter of a wet sclerophyll forest, and Dr Ian Common has found a range of 54–252 larvae in a series of samples of 130 g of leaf litter, dry weight, from forest litter in southern New South Wales.

The impact of these moths on nutrient recycling is very significant. Their litter-feeding larvae largely prevent the long-term build-up of leaf litter and this reduces the risk of fire enormously. Curiously, although their critical importance has been recognised for some time, no studies of these moths in leaf litter have been funded.

The contribution of moth larvae to the breakdown of the dry litter in sclerophyll forests in Australia has been greatly underestimated because techniques developed in Europe to extract organisms from litter have been used. The moist, deciduous litter in Europe is very different to the dry Australian litter and different organisms, behaving differently, are important in Australia. In Europe, as the litter dries out the organisms migrate deeper into the litter where it is moist, whereas

Australian organisms remain in place and are therefore not sampled using European techniques. Eucalypt plantations in Europe develop deep litter accumulations because the European fauna that breaks leaves down does not attack eucalypt litter.

While the moths' larvae are a most important natural control of leaf litter build-up, in a fire all are killed. Prescribed burning is designed to control the build-up of leaf litter yet it kills the natural litter controlling fauna. There are places in the southern tablelands of New South Wales that have not been burnt for more than 50 years yet have no excessive litter build-up, so the natural processes work if they are given a chance. So does prescribed burning do more harm, long term, than good? This whole question is desperately in need of

The larva of a moth of the genus *Garrha* (Oecophoridae) has come partly out of its case, made of two cut eucalypt leaves silked together. These larvae live on the driest litter and in hot conditions may tie the case to an upright stem to minimise the heat from the sun. Photo: Ian Common

comprehensive research and the role of moths can no longer be ignored.

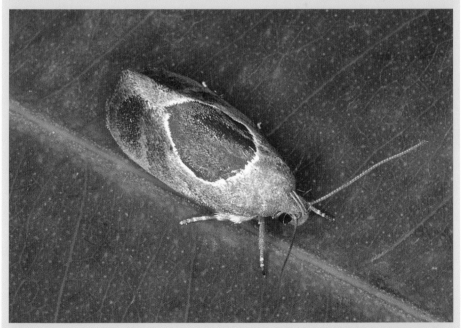

This common moth, *Hoplomorpha abalienella* (Oecophoridae), is found in eucalypt forests from the Atherton Tableland to eastern Victoria. The larva lives in curled dead leaves of eucalypts on the forest floor upon which it feeds. At rest in a dark place, the moth looks like it has a hole through it.

Xyloryctidae

- small to large
- usually smooth head
- wings held roof-wise
- antennae simple or pectinate (feathery) in male, held back along leading edge of wing
- palpi upturned, sickle-shaped, sharp-pointed
- hindwing broadly lanceolate

This is a diverse and widely distributed family in Australia. Most are small moths but many are giants with wingspans up to 75 mm.

The larvae of many species tunnel in stems, and some form silken tunnels in lichens or in clusters of leaves. The genus *Cryptophasa* contains many large species that typically bore in the trunks of trees. They leave their tunnel to cut leaves from the tree, which they drag back and tie with silk to the entrance to the tunnel and then feed from the shelter of the tunnel for the next few days.

There are about 250 species in Australia and 1200 worldwide.

This medium-sized moth, a species of *Eschatura*, is found in rainforest from Cape York to the Cairns district, Queensland. A very similar species is found in southern Queensland. The larvae are borers in rainforest trees, *Syzygium floribunda* (Myrtaceae) and *Elaeocarpus angustifolia* (Elaeocarpaceae). The reason for the lengthened wingtips is unknown.

The small *Catoryctis eugramma* is found in moist euca-lypt forests from the Atherton Tableland to southern New South Wales. The larva bores in the branches of *Casuarina* (Casuarinaceae). The bold longitudinal lines are typical of moths that rest in grasses or, in this case, the long needle-like branchlets of *Casuarina*.

The small *Telecrates basileia* is found in eucalypt forest and woodland in the top end of the Northern Territory and from Cape York Peninsula to Brisbane, Queensland. It lives in eucalypt forest and woodland. Nothing is known of its biology but a closely related species has larvae that bore in eucalypts.

This small, striking moth, a species of *Scieropepla*, is found in Western Australia from Busselton to Albany. Its biology is unknown but the closely related *S. trinervis* from south-eastern Australia has larvae that bore in the flower cones of *Banksia* (Proteaceae).

The small to medium-sized *Xylorycta ophiogramma* is found in tropical eucalypt woodlands from Cape York to the Atherton Tableland, Queensland. Nothing is known of its biology.

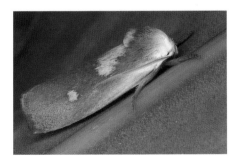

Cryptophasa rubescens is a large, common moth found from Coen, Queensland to Melbourne, Victoria. The larvae bore in the stems of many species of *Acacia* and cut leaves that they tie with silk to webbing at the entrance to their hole. They then feed on these leaves as they dry out. This is a male.

Cryptophasa xylomima is a large, common moth, found in the rainforests of northern Queensland from Cape York south to near Townsville. Nothing is known of its biology. The resting adult has the appearance of a broken piece of rotting branch.

Hypertrophidae

- small
- smooth or woolly head
- wings held steeply roof-wise
- antennae held back along leading edge of wing
- palpi upturned, sickle-shaped, sharp-pointed
- hindwing broadly lanceolate

Hypertrophids are found throughout Australia. Some species have shining metallic scales, often forming tiny eye-spots on the edge of the wings; others have silver wings or pale green wings that fade.

The larvae live in silken shelters on leaves. The pupa, like that of the Depressariidae, stands erect without a cocoon on its broadened base; this relates them to the Elachistidae, which also stand without a cocoon. The family is restricted to Australia and New Guinea, with about 50 recorded species.

Eupselia satrapella is a bright, little moth found in eucalypt forests from southern Queensland to Victoria. The larvae feed on eucalypts, starting with a mine in the leaf and later enlarging this to form a tunnel, open at one end, from which the larva emerges to feed on the surrounding leaf.

This moth is one of a pair of species that are indistinguishable in photographs. *Thudaca haplonota* is found from Geraldton to Cape Arid National Park, Western Australia; *Thudaca crypsidesma* is found in the west too, but also in South Australia, Victoria and Tasmania. The larvae of a related species feed on *Leptospermum* (Myrtaceae).

Hypertropha chlaenota is found from Rockhampton, Queensland through the moister parts of southern mainland Australia to Perth, Western Australia. It has been reared from a pupa found under the bark of a eucalypt that is undoubtedly the foodplant.

Depressariidae

- small
- smooth head
- wings usually held steeply roof-wise at rest
- antennae simple, held back along leading edge of wing at rest
- palpi upturned, sickle-shaped, sharp-pointed
- hindwings broad, lanceolate

This family of moths is found throughout Australia and cannot be easily distinguished from the Hypertrophidae without dissection. They also have a pupa that stands erect on a broadened base without a cocoon. There is a great variety in colours, patterns and shapes.

The larvae feed on the green leaves of plants forming a silken shelter, but a few feed on leaf litter. The pupae are found on leaves, stems or under bark. There are 58 recorded species in Australia and about 600 worldwide.

Pedois lewinella is a common species adapted for resting on the bark of trees. It is found from southern Queensland to Tasmania. The caterpillar is unknown, but the squat pupa has been found, without a cocoon, under the bark of eucalypts upon which the caterpillar probably feeds.

Tonica effractella is found from the Kimberleys, Western Australia, through the top end of the Northern Territory, and from Cape York to Proserpine in Queensland. The larvae bore in plants of the hibiscus and kurrajong families, including cotton. The moths are mostly found in monsoon forest.

This moth is found from the Atherton Tableland to near Mackay, Queensland. Nothing is known of its biology. The leading edge of the forewing has an unusual shape. The purpose of this is not known.

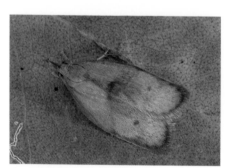

This small moth, a species of *Enchocrates*, is found in the south-western corner of Western Australia. Nothing is known of its biology.

Barantola panarista is found in rainforest from near Cooktown, Queensland south to Dorrigo, New South Wales. The moths in this genus are distinctive with long antennae and a characteristic wing pattern. Nothing is known of their biology. This is the only silver species; others are brown or grey. Moths with a similar wing pattern, belonging in the genus *Filinota* and possibly closely related, are found in Central America.

Elachistidae

- tiny
- rough-scaled head
- antennae held back along leading edge of wing at rest
- palpi upturned, sickle-shaped, sharp-pointed
- both wings narrow, pointed, with long scale fringes from the trailing edge
- usually grey

Elachistids are retiring moths that are rarely seen but are often active during the day in the coolest months.

The larvae mine in the leaves of grasses and sedges, starting with a long narrow mine that eventually widens. They leave the mine to pupate and the pupa is usually not protected by a cocoon. Most species are found in the cooler, moister parts of south-eastern Australia.

These moths have not been studied; currently 16 species are known but work in progress will bring the known Australian fauna to 140 species, with many still to be discovered. Worldwide about 250 species are known.

This very small moth, a species of *Elachista*, was reared from the sedge *Ficinia* (Cyperaceae) at Adelaide, South Australia. Nothing else is known about it. Photo: Lauri Kaila

This very small moth, a species of *Elachista*, was reared from the sedge *Carex appressa* (Cyperaceae) at Canberra, Australian Capital Territory. Nothing else is known about it. Photo: Lauri Kaila

Ethmiidae

- small to medium-sized
- head rough-scaled
- wings held roof-wise
- antennae held back along the leading edge of the wing
- palpi upturned, sickle-shaped, sharp-pointed
- hindwing broad, lanceolate, often black and white, or black and yellow
- forewing grey or white with numerous black spots

In Australia, these moths are mostly found in rainforest, but a few species are found commonly in the arid zone.

The larvae feed on plants of the Boraginaceae family, with the rainforest tree *Ehretia acuminata* supporting several species. They feed on green leaves under a web of silk or in silken webs on flowers. Some of them bore into dead and soft wood to pupate.

There are 14 species in Australia and more than 300 worldwide.

This moth is difficult to identify from a photograph but it may be *Ethmia thoraea*, which has bright orange hindwings, here completely hidden. It is found from the Atherton Tableland and Forty Mile Scrub, Queensland, south to Kiama, New South Wales and into the semi-arid zone as far as Cunnamulla.

Ethmia clytodoxa is a small moth found in rainforest from the Atherton Tableland, Queensland to Kiama, New South Wales. The larvae have been reared from silken shelters on the leaves of *Ehretia acuminata* (Boraginaceae) upon which they feed. Photo: John Stockard

Ethmia heptasema is a small moth found in rainforest from just north of Brisbane, Queensland to Narooma, New South Wales. The larvae have been reared from *Ehretia acuminata* (Boraginaceae), a common rainforest tree. Photo: John Stockard

Blastobasidae

- small
- head smooth-scaled
- wings held low roof-wise back along body at rest
- antennae held back along leading edge of wing
- palpi upturned, sickle-shaped, sharp-pointed
- hindwing narrow, lanceolate, with long scale fringes from trailing edge
- very drab grey

Blastobasids are found usually in the warmer and wetter parts of Australia with none in the arid zone.

The larvae are plant detritus feeders, preferring moist decaying plant material.

They have been found in the old male cones of kauri pines and in the old flower spikes and young fruit of palms. There are about 10 recorded species in Australia and about 300 worldwide.

This species of *Blastobasis* is found in southern New South Wales and is typical of the drab and consistent wing pattern found in this family. Although the family is relatively small in Australia, some species are very common.

This inconspicuous male, a species of *Blastobasis*, belongs to a group widely distributed, particularly in northern Australia. The larvae feed on dead or decaying plant material or occasionally on flowers. The group is unstudied in Australia.

Cosmopterigidae

- small
- head smooth-scaled
- wings held roof-wise over body when at rest
- antennae simple, held back along leading edge of wing at rest
- palpi upturned, sickle-shaped, sharp-pointed
- both wings may be narrowly lanceolate or very narrow with long scale fringes from the trailing edge of the wing

In Australia, gelechioid species of doubtful family are placed in the Cosmopterigidae, which is thus a diverse 'dumping ground' for many disparate groups of moths until their true relationships are understood.

The genus *Cosmopterix* contains tiny, very narrow-winged, elegant moths whose larvae mine in grasses. The very large genus *Macrobathra*, whose adults are more broadly-winged, have larvae that tie leaves of many *Acacia* (Mimosaceae) or *Senna* (Caesalpiniaceae) species. The larvae of some other species are scavengers or detri-

tus feeders, with one species eating the carton of paper wasps' nests. Some species have adults with very narrow wings and red eyes. A few have gall-forming larvae or feed in the hidden epicormic buds of eucalypts. One group lives in leaves, dead or alive, of screw palms, *Pandanus* (Pandanaceae), cutting small cases from the leaves giving them a shot-holed appearance.

Australia has a rich fauna with about 400 recorded species in the family but a great many unrecorded. Worldwide there are about 1600 species.

This tiny but exquisitely patterned moth belongs to an almost cosmopolitan genus, *Cosmopterix*, comprised of many species. All the species look quite similar. The larvae of species overseas mine in the blades of grasses, sedges and sometimes other plants.

The genus *Leptozestis* occurs throughout Australia and contains many species. They are characterised by numerous tufts of erect scales on their narrow wings. The few species that have been reared have been reared from galls on trees, borers in dormant buds and from seed cones.

Morphotica mirifica is a brilliant little moth found in the top end of the Northern Territory and on Cape York Peninsula. Its resemblance to species of *Macrobathra* would suggest that *Acacia* would be a starting point to look for the early stages.

Australia has numerous very similar species in the genus *Macrobathra;* most are difficult to tell apart. The genus is found throughout Australia. In arid areas, mulga supports a varied fauna of these moths. The larvae feed on *Acacia* leaves, tying them together with silk.

Labdia ceraunia is so small that its brilliant colours are hardly visible without magnification. It is found in rain-forest from Kuranda to Brisbane, Queensland. The red eyes and metallic scaling on the wings are typical of the group of moths to which it belongs.

Gelechiidae

- small
- smooth-scaled, slightly narrow head
- wings held flat or low or high roof-wise over body
- antennae simple, held back along leading edge of wing at rest
- palpi upturned, sickle-shaped, sharp-pointed
- hindwings narrow or broad but with a sinuate margin caused by an extension of the apex of the wing; when the hindwings are narrow then there are long fringes of scales on the trailing edge and the apex of the wing greatly extended, making a very sinuate margin
- adults sometimes run very fast and straight on the ground

This is a very large family well represented in Australia where the mostly Australian *Ardozyga* group is dominant. This group has hindwings less sinuate and broader than other species.

Gelechiid larvae are often reddish in colour and have very diverse habits: some tie leaves with silk, some bore in fruit or flowers, others form galls or bore in stems or roots. The larvae pupate in the galleries or shelters formed by the larvae. The family contains some serious pests such as the potato moth (*Phthorimaea operculella*), pink bollworm (*Pectinophora gossypiella*) and the Angoumois grain moth (*Sitotroga cerealella*). There are about 800 recorded species in Australia but at least as many are unrecorded. Worldwide there are about 4500 species.

Dichomeris ochreoviridella is found in rainforest or monsoon forest in Queensland, from Cape York to Gympie. It also occurs in New Guinea. Nothing is known of its biology. The fine sickle-shaped labial palps, characteristic of the gelechioids, are very evident here.

Ardozyga chionoprora is found in wet eucalypt forest from southern Queensland to western Victoria and Tasmania. The adult moth rests during the day on the bark of trees. The caterpillar of this species is unknown, but related species tie together two eucalypt leaves and live in the shelter between the leaves.

Ardozyga nyctias is found throughout mainland Australia except for the arid areas. Nothing is known of its early stages.

Ardozyga stratifera is found through the moister districts of mainland Australia south of Brisbane, Queensland to Perth, Western Australia. The larva joins two leaves with silk and lives and feeds in this shelter. It is found on eucalypts.

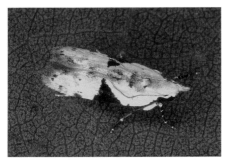

Anarsia anisodonta is a neat little moth found in New Guinea and from Cape York to the Cairns district and Atherton Tableland, Queensland. Nothing is known of its biology but some other Australian species in the genus feed on *Acacia* as larvae.

This specimen of *Hypatima discissa* may be a faintly marked individual of this species. It is found in monsoon forest from Cooktown to Cairns, Queensland. Nothing is known of its biology.

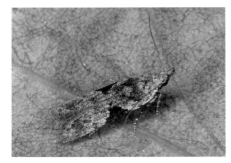

Anarsia molybdota is found throughout Australia south of the Tropic of Capricorn, particularly in semi-arid regions. It is absent from Tasmania. The caterpillar joins two phyllodes of *Acacia* with silk and lives in the shelter so formed.

This very small moth is probably in the genus *Thiotricha*, or a closely related genus. The species is found on the Atherton Tableland, Queensland.

This *Helcystogramma* species is found in wet places in central and southern New South Wales and eastern Victoria. It is sometimes active in the daytime but also comes to light. The larvae of some related species feed on grass and silk together the margins of the leaf blade, towards the tip, and live in the hollow cylindrical shelter so formed. From their shelter, they feed on the grass blade towards the tip, and then towards the base, leaving the mid-rib that bends with the weight of caterpillar and shelter.

Hypatima spathota is very widespread, found from Sri Lanka, Nepal and India eastwards to Australia. In Australia it is found from Iron Range south to just north of Brisbane, Queensland. The larvae have been reared overseas from white cedar (*Melia azedarach*, Meliaceae) and mango (*Mangifera indica*, Anarcardiaceae).

Chaliniastis astrapaea occurs in rainforest from Cooktown, Queensland to Nowra, New South Wales. The larvae have been reared from the seeds of *Euroschinus falcata* (Anarcardiaceae). The classical gelechioid palpi are very clearly shown. This may be a case of a close foodplant association as the foodplant is recorded from Cooktown to Jervis Bay, a remarkably close match to the range of the moth.

Lecithoceridae

- small
- smooth head
- wings held flat back over body at rest, or sometimes extended out flat but held up off the substrate
- antennae long (longer than wings), sometimes thickened, pointed forward in front of head at rest
- palpi upturned, sickle-shaped, sharp-pointed
- hindwings broad with short scale fringes

These moths occur widely in Australia but mainly in tropical and subtropical areas and some species are very common. The larvae are unusual in the gelechioids in being very hairy and some are detritus feeders, having been reared from leaf litter on the forest floor.

The group has not been studied in Australia, where there are about 50 recorded species. The family is most prevalent in the African and Oriental regions where about 500 species are known.

This species of *Lecithocera* is found on the Atherton Tableland and Cairns district in Queensland. The larvae of related species feed on leaf litter on the ground. This moth clearly resembles some object that would be avoided by predators, perhaps a thorn.

Crocanthes sidonia is found in New Guinea and in Australia from Cape York to the Cairns area, Queensland. Its early stages are unknown. This is a male.

Crocanthes characotis is found in the top end of the Northern Territory and from Cooktown to Tully, Queensland. Nothing is known of its biology but it has been collected in rainforest.

Scythrididae

- very small
- smooth head
- wings held roof-wise over body at rest
- antennae held back along leading edge of wing
- palpi upturned, sickle-shaped, sharp-pointed
- both wings narrow, hindwing with long fringe of scales on trailing edge
- body somewhat short and stout

Moths in this family are often black, or black with orange markings, but a few are grey. Some species are mostly day-active and visit flowers but may be collected at light, particularly if the light is near flowers.

The larvae of one species have been found feeding on the leaves of grasses.

Moths of the genus *Paratheta* fly at night and have been reared from galls on the twigs of *Sclerolaena birchii* (Chenopodiaceae).

There are 24 species recorded in Australia and about 700 worldwide.

Eretmocera chrysias is found from north-western Australia to Maryborough, Queensland. It often visits small flowers in the day but may also be found at night, particularly if the light is near flowers. The larvae have been reared from the grass *Enteropogon acicularis* (Poaceae). This is a female.

Cossidae (Wood Moths)

- large to very large
- rough-scaled head; small compared to body
- wings held roof-wise
- male antennae very short, pectinate, pectinations rapidly shortening halfway along, rarely long all the way to tip
- antennae held back under forewings at rest
- wings narrow for the size of the moth
- palpi very small
- proboscis greatly reduced

Most cossids are brown, grey or white, often finely speckled with a net-like pattern and with a characteristic black and iridescent blue inverted 'V' on the thorax. The eggs may be large and placed in small batches, or small and deposited in large numbers in crevices in trees and covered with a glutinous secretion that later hardens.

The young larvae, on hatching, often disperse on strands of silk carried by the

This large cossid, a species of *Endoxyla*, is found in open tropical eucalypt woodland on Cape York Peninsula and the top end of the Northern Territory as far west as Keep River. Nothing is known of its biology. This is a male.

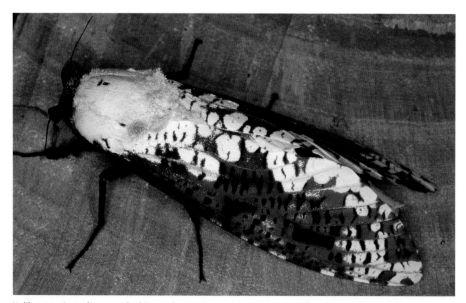

Unlike most Australian cossids, this very large species, *Xyleutes persona*, is found in rainforest and is widely distributed overseas from India to New Guinea. In Australia it is found from Cape York south nearly to Townsville, Queensland. Overseas the caterpillar has been found boring in a range of trees mostly from the family Caesalpiniaceae. While conspicuous in this photograph, if it were resting on rotten wood and white fungus it would be well camouflaged. This is a male.

Rather than bore deep in the wood like the other cossids shown here, the larva of *Culama australis* bores just beneath the bark of *Eucalyptus*. It is very widespread in eucalypt forests and mallee from southern Queensland to Tasmania and west to Western Australia. It is very easily confused with other similar moths. This is a male.

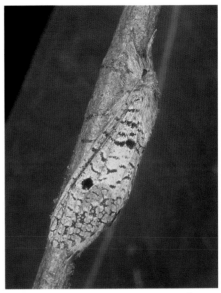

Endoxyla encalypti flies in midsummer and is common from southern Queensland to eastern Victoria. This very large moth is frequently seen in the suburbs of Canberra. The larvae are gigantic and bore in the butt and upper roots of living *Acacia* trees. The empty pupal cases are often seen protruding from the butt of the tree after the moth has flown away. Another smaller, greyer species is found boring in the trunks and upper branches of *Acacia*. This is a male.

This large cossid, *Endoxyla epicycla*, or very similar ones that may be the same species, is found from northern Queensland to the mallee of South Australia. This female has a long protruding ovipositor, but the foodplant of the caterpillar is not known. Like many other *Endoxyla* it will be a borer in living wood.

wind. The larvae are borers in stems, branches, trunks, butts or roots of trees or shrubs. They usually bore singly but a few are communal. In most cases, they bore into the centre of the trunk to provide a shelter, and the larva feeds in an entrance vestibule intersecting the bark and sapwood. As the plant tries to repair the tissue the larva continues to graze, remaining inside the tree with only a small hole to the outside to expel droppings.

Larvae normally bore upwards and pupate in the tunnel after scoring the bark at the entrance to aid the escape of the moth. Even for the giant species, larval life is rarely more than two or three years.

The pupal shell protrudes from the trunk after the emergence of the moth.

There are 86 described species found throughout Australia but at least another 100 are known – some of them very large moths. There are about 670 species world-wide.

The Witjuti Grub

Witjuti grubs are one of the best known food sources for desert Aboriginal people. The name is used in many different senses in Australian English. In the widest sense, witjuti grub means just about any large insect larva that might be eaten by humans and could be the larva of a moth or a beetle. In its strict sense the word has a much more precise meaning, which was investigated by the lepidopterist-anthropologist Norman Tindale in the 1950s. The term 'witjuti' is strictly the name of the shrub or small tree *Acacia kempeana* that grows widely in central and western Australia. The Arabana people knew the larva as 'mako witjuti' with 'mako' meaning 'grub' and 'witjuti' the tree. Most of Tindale's work,

however, was done at Ooldea in South Australia, where larvae were dug from the roots of *Acacia ligulata* and known by the Ngalea people as 'mako wardaruka', with 'wardaruka' meaning the *Acacia ligulata*. Tindale reared moths from Ooldea and identified them as *Endoxyla leucomochla* that is widespread in arid areas across southern Australia. However, no moths have ever been reared from the witjuti bush of central Australia, and it is by no means certain that it is the same moth as that at Ooldea or that only one moth is involved. Nor is it certain whether the moths reared from *Acacia ligulata* by Tindale at Ooldea were really *E. leucomochla*.

The cossid moths of central Australia are

This is the larva of one of the large species of *Endoxyla*, (Cossidae). They bore in the trunk and large branches and were sometimes cut from the timber by aboriginal Australians.

very poorly known but at least seven large species have been collected and more may be present. So far, the real *E. leucomochla* has not been found there.

Today the witjuti grub is a household word, larvae are constantly dug for tourists and similar cossid grubs are marketed for human consumption. But scientifically they are unstudied and almost unknown.

Tindale also mentioned several other cossid larvae from the roots of chenopodiaceous shrubs that were also dug for food in central and South Australia. F. P. Dodd, at the beginning of the last century, described the local Aboriginal people cutting cossid larvae from the trunks and branches of eucalypts on the Atherton Tableland but also said that in spite of being plentiful, cossid larvae were

unknown as a source of food in the Darwin area. There is no doubt that several of the large eucalypt-feeding cossid species were eaten on the Atherton Tableland.

The term 'bardi grub' has a similar history. It was first applied to beetle larvae taken from dead grass tree (*Xanthorrhoea*) stems at Albany in Western Australia but now has come to be applied to many edible larvae. In one sense it is applied to larvae of the hepialids or swift moths that bore in the soil beneath river red gums and feed externally on the roots. These larvae may be eaten directly or used for fishing. The hepialids involved are probably *Trictena atripalpis* and *Abantiades marcidus*, but again this has never been investigated.

This is a cossid moth of the genus *Endoxyla* to which the witjuti grub belongs. The larva of this moth bores in eucalypts in south-western Australia.

Cossids and cockatoos

Most people in south-eastern Australia will have seen chunks of wood torn from trees and lying on the ground. This is the work of yellow-tailed black cockatoos that have torn the wood up to obtain the larvae of cossids, xyloryctids or sometimes hepialids. The cockatoos need protein for nestlings, and the usual diet of eucalypt seed, banksia seed or pine nuts needs supplementing to provide enough protein.

The cockatoos search the forest in groups and, when looking for larvae, they are guided by subtle signs like swelling in the trunk, or silk and droppings, or sawdust that covers over the entrance hole. When they find an infested tree they sink their beaks into the wood and if they detect movement there may be a larva present. On the vertical trunks of big trees the cockatoo rips several large pieces of bark from well above the boring and pulls them down to below the boring but does not cut them off from below. The cockatoo then climbs onto these pieces of bark and uses them as a springboard to provide a little extra leverage while it extracts the larva. The bird then tears out strips of wood until the larva is exposed. Normally this causes little damage to the forest but in a plantation for paper pulp it can be more serious. The cossids' damage may slow the growth of the trees but when infesta-

These plantation trees of *Eucalyptus grandis* have been damaged by yellow-tailed black cockatoos feeding on cossid larvae. A strong wind has resulted in their fall. Photo: Ray McInnes and John Green

tions are high and cockatoos active then the wind-throw of trees in high winds may lead to great loss of trees.

Ray McInnes from CSIRO Entomology found that in trees harvested while still young and while cossid damage was still close to the ground, damage could be averted by allowing a shrubby understorey to grow. The cockatoos possibly feared terrestrial predators and in shrubby situations they could not see any lurking danger and did not come low to eat the cossids. The forests could also be protected to some extent from wind-throw by letting the understorey grow.

Yellow-tailed black cockatoo. Photo: Michael Todd

This sapling of *Eucalyptus grandis* has been opened by a yellow-tailed black cockatoo to extract the cossid larva. Photo: Ray McInnes and John Green

This billet has been split open to show a cossid larva, probably *Endoxyla cinereus*, in its tunnel. This one is fully grown and has webbing and repellent chemicals in place to deter entry to the tunnel while it is a helpless pupa. Photo: Ray McInnes and John Green

Dudgeoneidae

- large
- rough-scaled head
- wings held roof-wise
- antennae of males have a single row of very short pectinations
- antennae held back along leading edge of wing at rest
- forewings chocolate brown with silver spangles

Unlike cossids, these moths have hearing organs at the base of the abdomen. They have a characteristic wing pattern unlike any other moth, with rich brown or chocolate brown forewings and two large patches of silver spangles.

The larvae bore in the branches of *Canthium* but have not been studied for many years. The pupal shell is protruded before emergence. The adults come to light.

Three species have been described from Queensland, mostly in tropical rainforest, and several other species are known. Overseas, the family Dudgeoneidae is known from Africa, Madagascar, India, South-East Asia and New Guinea with only about six species.

Dudgeonea actinias is a medium-sized moth found from Cape York Peninsula south to Townsville, Queensland. The larvae have been found boring in the branches of *Canthium* (Rubiaceae). There are other species in the genus in Africa, India, South-East Asia and New Guinea.

Tortricidae (Bell Moths, Leaf Rollers)

- small to medium-sized
- head usually smooth-scaled
- wings held flat roof-wise or rolled about body
- antennae simple, held back along leading edge of the wing
- forewing near base often strongly curved
- palpi projected forward directly in front of head
- both wings broad with short scale fringes
- hindwing often slightly pointed near front edge

The family Tortricidae contains the codling moth and oriental fruit moth, both serious pests, particularly of fruit. Another serious pest of fruit, *Epiphyas postvittana*, the light-brown apple moth, is a native Australian that has become a pest in several overseas countries. Most adult tortricids come readily to light but there are a small number of diurnal species.

The larvae are typically 'leaf-tiers': they silk leaves together and live and feed in the shelter so formed. Some are borers in stems, cones, under bark, in seeds, and many are borers in fruit. One group feeds on dead leaves on the forest floor. Most live solitarily but a few live in large communal webs. They pupate in the larval shelters from which the pupal shell protrudes when the moth emerges.

The family is found worldwide with more than 5000 species; there are more than 1200 species in Australia.

This small moth, a species of *Trymalitis*, is common in rainforest from Iron Range south to Tully, Queensland. Nothing is known of its biology. The larvae of a very similar species in Queensland have been reared from *Mimusops elengi* (Sapotaceae). A similar but larger species is found in southern Queensland and northern New South Wales.

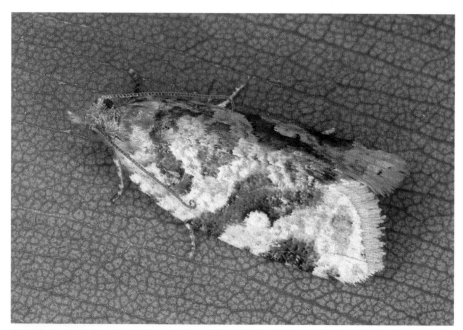

Epitymbia eudrosa is found mainly in wet or dry rainforests in New Guinea and in Queensland as far south as Rockhampton. It has been reared in the laboratory but in nature the larvae probably eat dead leaves on the ground (but probably not eucalypt leaves).

This bright little moth, *Ophiorrhabda phaeosigma*, is found in the Kimberleys, Western Australia and the top end of the Northern Territory and in northern Queensland from near Cape York to Chillagoe. It is most plentiful in the moister tropical open forests. The larvae have been reared from the fruit of *Syzygium armstrongii* and *S. eucalyptoides* (Myrtaceae).

Proselena tenella, small. This moth is found in the tablelands of central and southern New South Wales and near Adelaide, South Australia. It is associated with blackthorn (*Bursaria* spp., Pittosporaceae) where the larva at first mines in the leaf, later using the leaf as a case while feeding on the leaves.

This unusual and boldly marked moth, *Goboea copiosana*, is found from Noosa and the Bunya Mountains, Queensland to Broken Head, New South Wales. The larvae have been reared webbing the leaves of *Araucaria* spp. (Araucariaceae), hoop pine and bunya pine.

Dudua aprobola is recorded from Africa and the Seychelles through tropical Asia and China to Tahiti. In Australia it is found from south of Broome to Derby, Western Australia and in the top end of the Northern Territory and in Queensland from Cape York to Burleigh Heads. The larvae feed on the young leaves and sometimes flowers of a very wide range of plants.

Ancylis artifica is found in monsoon forest and rainforest margins in the top end of the Northern Territory and from north of Cooktown, Queensland to Grafton, New South Wales. The larvae have been found tying the leaves of *Alphitonia excelsa* (Rhamnaceae) together with silk and feeding on the leaves. This is a male.

Sycacantha exedra is found in rainforest in New Guinea and from Cape York to Maryborough, Queensland. Moths must be handled carefully as critical features such as the tuft of scales so evident here are easily rubbed off. Nothing is known of its biology. This is a male.

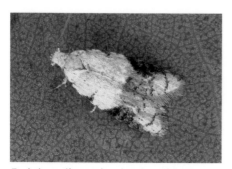

Tracholena sulfurosa is found on the tablelands and western slopes of southern Queensland and New South Wales wherever cypress pines are found. The larva has been found tunnelling in the bark of garden pines (*Cupressus*) and no doubt does the same on native *Callitris* (Cupressaceae).

This small moth, a species of *Holocola*, is found in south-western Western Australia. It is one of many species in a typically Australian group that feed on Myrtaceae. Nothing is known of the biology of this species.

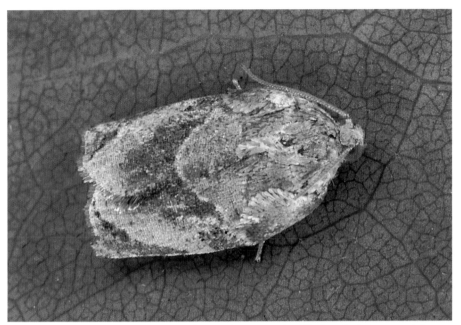

Homona spargotis is found from Mossman to Shute Harbour, Queensland. The larva feeds on many rainforest plants and is a pest of many fruit trees. The fringe of scales on the wing beneath the tip of the antenna marks a fold of the wing membrane that protects specialised scent scales. This is a male.

This medium-sized moth, a species of *Cryptoptila*, is well known in rainforest from the Atherton Tableland, Queensland to Wollongong, New South Wales. The larvae have been reared from leaves of *Pennantia cuninghamii* (Icacinaceae).

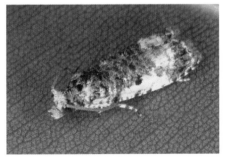

Phricanthes asperana is found on the margins of rainforest or wet eucalypt forest from Mossman in northern Queensland to Batemans Bay in southern New South Wales. The larva folds a leaf of *Hibbertia scandens* (Dilleniaceae) over to form a roomy shelter in which it hides and feeds.

Castniidae (Sun Moths)

- large
- smooth-scaled head
- wings held flat while basking, sometimes up over the back, but when at rest steep roof-wise clasping substrate
- antennae held out and to side when basking, back along leading edge of wing at rest
- antennae simple but thickened with well-developed and prominent club at the tip
- female with long needle-like egg-laying tube
- strictly day-flying

These are the only moths that have such abrupt, prominently clubbed antennae like those of butterflies. They are strictly diurnal and often fly only in sunshine during the hottest hours of the day. They bask, and they recognise mates visually. The adults rest with wings extended and body pressed to the ground while basking but otherwise they rest with wings held roof-wise.

Most species have cryptic grey or brown forewings but are black or brown with either red or yellow spots on the hindwing.

This large day-flying moth, *Synemon plana*, is found in native grasslands from Bathurst, New South Wales to the Victoria–South Australia border. This habitat has been largely destroyed for agriculture and this moth is one of the flagship species that draws attention to the need for grassland conservation. The larvae feed underground on roots and stems of *Austrodanthonia* (wallaby grass). This is a male. Photo: Ted Edwards

This large moth, *Synemon magnifica*, is found from near Toowoomba, Queensland to near Nowra, New South Wales. It flies very rapidly in hot sunshine in midsummer over flat sandstone rocks around the edges of which the larval foodplant *Lepidosperma viscidum* (Cyperaceae) grows. The caterpillars feed on the underground rhizomes of this sedge. This is a male. Photo: Michael Braby

The larvae feed underground on the roots and rhizomes of grasses and sedges. The fully grown larva may make a tunnel to the soil surface or prepare a silken dome from which the moth will emerge. The pupal shell protrudes from the shelter when the moth emerges. Some Australian species inhabit endangered native grasslands.

Castniids are found in Central and South America with a few in South-East Asia; those in Australia are said to be most closely related to the South American ones. There are 24 described species distributed throughout Australia (with as many more currently being studied) but they are absent from Tasmania. There are about 90 species worldwide.

Synemon wulwulam is a large day-flying moth found in semi-arid areas from Cloncurry, Queensland to Derby, Western Australia. The larvae are thought to feed on the underground rhizomes of the grass *Chrysopogon*. Photo: David Rentz

Brachodidae

- small
- smooth-scaled, broad head
- wings held low roof-wise at rest
- antennae held forward of head when at rest
- thorax very stout
- wings relatively short

These moths are small and have short antennae, usually pectinate or with very short pectinations in males. The labial palpi are small, the proboscis is unscaled, and nearly all have short wings and a very stout thorax.

There are three groups in the family; one is a southern group containing the genus *Miscera* that is mostly diurnal, often flying among sedges or grasses during the morning but sometimes coming to light. A northern group containing *Synechodes* and another genus comes to light at night, and several have been reared from larvae boring in the stems of palm flowers and leaves or from fruit of lawyer palms. The third group, represented in Australia by two rarely seen species of *Nigilgia*, is diurnal and visits flowers in rainforest.

This is a small family of about 100 species, found worldwide (except for North America) and there are 24 species in Australia.

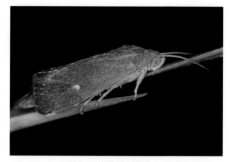

Miscera centropis is a small, day-flying moth found in sedgelands of Western Australia in the Perth area. Nothing is known of its biology. Photo: Axel Kallies

The female *Synechodes coniophora* is different from the male: it has red banding on its wings. Both sexes of the moth come to light. Photo: Axel Kallies

Synechodes coniophora is found in northern Queensland from Cape Tribulation to Kuranda. The larva bores in the stems of the flower spikes and leaves of palms. This is a male. Photo: Axel Kallies

Sesiidae (Clearwing Moths)

- small to medium-sized
- smooth head
- wings held flat back over body at rest
- antennae simple, thickened, gradually broadening to two-thirds of length where broadest
- antennae held out wide in front of head
- wings narrow, usually with large clear areas without scales
- body very long, wasp-like
- long stout legs

Sesiid moths are most distinctive. They look like wasps with narrow clear wings and a large banded body. They have prominent ocelli, the proboscis is unscaled and the palpi are directed forward and slightly curved upwards. They are diurnal and will visit flowers.

The larvae tunnel in the trunks and stems of trees, shrubs and vines, and the pupa protrudes from the tunnel when the moth emerges.

Most Australian species inhabit rainforest but they are very rarely collected. One species, often found boring in the galls of the native cherry, *Exocarpos cupressiformis* (Santalaceae), is found as far south as Melbourne. Overseas studies have been advanced enormously by using sex pheromones to attract the adults and many more species may be discovered in Australia.

There are more than 1000 species worldwide; 16 are known from Australia.

The medium-sized *Ichneumenoptera chrysophanes* is found from Cairns, Queensland to Melbourne, Victoria. The larvae feed in galls on *Exocarpos cupressiformis* (Santalaceae) and in the bark of trees such as *Alphitonia* (Rhamnaceae), *Eucalyptus* (Myrtaceae) and *Ficus* (Moraceae). The moths are day-flying but are rarely seen. This is a male. Photo: Axel Kallies

The medium-sized *Nokona carulifera* is found on Cape York Peninsula and on the Atherton Tableland. It is day-flying and resembles a wasp. Nothing is known of its biology. This is a male. Photo: Axel Kallies

Ichneumenoptera chrysophanes looks very similar to a wasp. Because it is day-flying, a resemblance to day-active wasps is helpful in avoiding predation.

Choreutidae

- small
- smooth head
- wings held back over body roof-wise but out from the body, raised and with wings slightly bent
- antennae short, simple, held out in front
- both wings short, broad
- some walk with a 'dancing' gait
- often diurnal

These moths have prominent ocelli and the proboscis is scaled at the base. They have broad wings, often with metallic scales in the markings and rest with the wings in a characteristic pose.

The adults are active by day; they visit flowers and walk in a jerky manner. There are five distinct groups in Australia, one with larvae on fig trees occurs in rainforest north from Sydney, and another group feeds on the foliage of daisies in alpine areas of New South Wales, Victoria and Tasmania.

There are about 350 species worldwide, with about 15 in Australia.

Saptha libanota is mostly active in the daytime. It is found in rainforests from Cape York to Townsville, Queensland. The larvae and their foodplant are unknown.

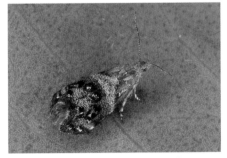

Tebenna micalis is a very wide-ranging and common moth found from Europe to the Pacific Islands, and in all states in Australia, but possibly not in the most arid areas. The larvae feed under a silken web on the underside of the leaves of daisies and thistles, eroding the leaf surface. The bending of the wings near the tips is characteristic of this moth.

Zygaenidae (Foresters)

- small
- smooth head
- wings held back over body low roof-wise
- antennae pectinate to the tip in males
- antennae held out at 45 degrees and in front of head
- short, somewhat stout bodies
- forewings shining green, copper or black or sometimes wasp-like in orange and black, or white and black
- mostly active by day

These are small moths, with an unscaled proboscis and small palpi. They have very large pincushion-like organs (chaetosemata) over the top of the head, almost meeting in the middle but only visible with a lens.

The adults are active by day, often feeding at flowers, but some also come to light. The adults produce cyanide to deter predators. The larvae are plump, hairy and feed on the leaves of many plants, particularly Dilleniaceae and Vitaceae (native grape)

where they erode the upper surface of the leaf, leaving characteristic tracks. The cocoon is spun in a shelter away from the foodplant, and the pupal shell protrudes after the moth has emerged.

All Australian species belong to the subfamily Procridinae, and many new species have been described recently from Australia bringing the number to 43, although more are known to occur. The family is cosmopolitan with about 1000 species.

Pollanisus nielseni is a brilliant shining moth that flies in the daytime. It is found in Western Australia, mostly in the Geraldton area, but also near Esperance. The larvae feed on the leaves of *Hibbertia spicata* (Dilleniaceae). In eastern Australia a brilliant green species, *Pollanisus viridipulverulenta*, is very common.

This moth, a species of *Pollanisus*, was photographed at Taree, New South Wales and may be found south to Batemans Bay, New South Wales. The larva was feeding on *Cissus antarctica* (Vitaceae), a native grapevine found in or near rainforest. Photo: John Stockard

Lacturidae

- small to medium-sized
- head smooth-scaled or rough
- wings held roof-wise over body at rest
- antennae simple, slightly thickened
- antennae held in front of head and out at 45 degrees
- forewings often brightly coloured in red and yellow, or red and black

These moths have prominent ocelli and large chaetosemata; the proboscis is unscaled and the labial palpi are small. The adults are very resistant to cyanide and probably produce it for protection.

The larvae are also brightly coloured and without obvious legs; they are slug-like and some look like blobs of jelly. They erode the leaves of the foodplant and pupate in a tough cocoon away from the leaves, with the pupal shell protruding on emergence.

This family is mostly found in tropical and subtropical rainforest but some occur in monsoon forest and tropical savannah. The larvae of *Eustixis caminaea* may be found in great numbers on the leaves of the Port Jackson Fig along the foreshores of Sydney Harbour.

There are 28 species known from Australia and more than 100 worldwide.

Eustixis leucophthalma is found in rainforest from the McIlwraith Range and Mossman, Queensland south to Grafton, New South Wales. Nothing is known of its biology.

Eustixis erythractis is found in monsoon forests and tropical eucalypt woodlands from the Kimberleys, Western Australia through the top end of the Northern Territory, and in Queensland from Cape York to Yeppoon. A small to medium-sized moth, it resembles many other species that often occur in the same areas. The early stages and foodplant are unknown.

This moth, a species of *Eustixis*, is found in rainforest only on the Atherton Tableland, Queensland. It is closely related to *E. sapotearum* that is found in rainforest in southern Queensland and in New South Wales, south to Wollongong. The larvae of the latter are brightly coloured, sticky and slug-like, with inflatable white lateral bags, and feed on *Pouteria australis* (Sapotaceae). This is a female.

Limacodidae (Cup Moths)

- medium-sized to large
- head rough-scaled, often small compared to body
- wings held steeply roof-wise over body
- antennae in male short, pectinate, with pectinations shortening rapidly towards tip
- antennae held back along leading edge of wing at rest
- wings short and very broad
- body short and stout
- body covered in long hair scales and wings covered in slightly 'fluffy' scales

These moths usually have no proboscis; their labial palpi are small and directed forwards.

The larvae are well known for their bright colours, flattened slug-like shape and the ability of some to sting if handled. They feed openly on the leaves of a wide range of foodplants. They pupate in impressively neat cocoons, almost spherical in shape, and spun on twigs. Some pupate in the litter, also in almost spherical cocoons. To emerge, the pupa pushes up a cap prepared by the larva, and the pupal shell is protruded on emergence. The unique cup shape of the cocoon minus cap gave rise to the common name.

About 70 species occur in Australia in all states; over 1000 species are known worldwide.

Comana albibasis is a common, medium-sized moth found from just north of Cooktown south to Brisbane, Queensland. Nothing is known of its early stages but the caterpillars of a related species have been reared from *Melaleuca*. Some other species in the genus are spectacularly coloured in red. This is a male.

Pseudanapaea denotata is a small moth, which is found from Iron Range to Brisbane, Queensland. Its biology is not known but a related species has a green larva, very humped at the front, almost jelly-like, but not spectacularly coloured like many other cup moths. It feeds on eucalypts.

This large moth, *Eloasa callidesma*, is found in wet eucalypt forest and rainforest from the Atherton Tableland to Toowoomba, Queensland. Nothing is known of its biology. This is a male.

The larvae of Limacodidae are all slug- or slater-shaped, and are often brightly coloured. Many have hairs that can cause a sharp sting if the larva is handled. They feed on a very wide range of plants.

The larva of *Doratifera vulnerans* feeds on many Myrtaceae, including *Eucalyptus*, *Lophostemon*, *Angophora* and *Melaleuca*, as well as on guava. It is found in all mainland states and is very common. This larva has been disturbed and the extended rosettes of stinging hairs are ready for action.

Epipyropidae

- small
- smooth head
- wings held steeply roof-wise, almost vertical
- head and eyes very small
- antennae very short, broadly pectinate in male, held back along leading edge of forewing at rest
- body small; very short and stout
- both wings short, very broad
- wings dull, dark grey
- adults often come to light at dusk

The moths have no proboscis and the labial palpi are minute. Epipyropids are crepuscular and are frequently the first moths to arrive at light after sunset.

The larvae are carnivorous and feed on the body juices of leafhoppers. The newly hatched larva is very active in finding a host, but once it commences feeding on the host it cannot transfer to another. The host dies after the larva has finished feeding and the larva lowers itself on silk to the ground to spin a white waxy cocoon. The pupal shell is protruded on emergence.

Most species occur in northern Australia but one reaches as far south as Canberra where it parasitises the passion-vine leafhopper. There are four species known from Australia and about 40 species in the world.

This small moth, a species of *Heteropsyche*, is one of a group of parasites of leafhoppers. They occur very widely in northern Australia but one species is found as far south as Canberra, Australian Capital Territory. This is a female.

Cyclotornidae

- small
- small, rough-scaled head
- wings held roof-wise
- antennae simple, slightly thickened, held out from head or back along wings at rest
- wings lanceolate
- thorax and body short and very stout

These are small moths with no proboscis; the labial palpi are short and drooping. They may be grey, sometimes with one or two black marks, or they may be straw-coloured. One has a bold pattern of black and yellow.

Cyclotornids are nocturnal and come to light. The larvae are carnivorous and parasitise leafhoppers. The larger species complete the first part of larval life on the leafhopper and then leave the leafhopper and spin an oval cocoon in which they moult into a flattened larva. This larva is then carried by a meat ant to its nest, where the larva feeds on the larvae and pupae of the ants. The larva produces an anal secretion sought by the ants, and as the larva is not armoured, it must also produce substances that prevent the ants attacking it. When fully grown, the larva leaves the ant nest to spin its cocoon in a sheltered place. The pupal shell protrudes from the cocoon after emergence. Larvae of the smaller species are thought to pupate in a flattened cocoon after leaving the leafhopper host.

The moths are found in eucalypt forests and arid areas, particularly in northern Australia, but some are found in most parts of mainland Australia. The family is confined to Australia, with five species named but more than a dozen known.

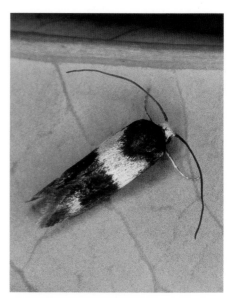

This small moth, a species of *Cyclotorna*, is found in moist eucalypt forests only in central and southern New South Wales and eastern Victoria. Most of the other known species are grey or straw yellow and occur in more arid habitats. The very broad thorax may be seen in the photograph.

Immidae

- small
- smooth head
- wings held flat back over body
- antennae simple
- antennae held back over wings, nearly together, at rest
- wings short, broad

These are small moths, usually with chaetosemata; they have a well-developed proboscis and the labial palpi are prominent and curve gently upwards. The wings are broad and the body is fairly stout. Some species have a yellow pattern on the forewings but some are grey, and a few have a hyaline hindwing with a black margin. One species in rainforest in northern Queensland is bright orange and black, but the reason for the bright colours is not known.

The larvae feed exposed on the leaves of the foodplant and drop on a thread of silk if disturbed. The loose cocoon is spun in a sheltered place and the pupal shell does not protrude from the cocoon after emergence.

There are 13 species known from Australia, most in northern Australian rainforests, and about 246 species in the warmer parts of the world.

Imma lyrifera has been found in rainforest from Cape York to Atherton, Queensland. Nothing is known of its biology.

Imma acosma is found on the western fringe of the Atherton Tableland, Queensland and from Yeppoon, Queensland south to Narooma, New South Wales. The larva feeds on the leaves of *Hymenanthera dentata* (Violaceae). This is a female. Photo: Ian Common

Copromorphidae

- small
- smooth head
- wings held flat back over body
- antennae slightly pectinate and flattened
- antennae held back along leading edge of wing at rest
- wings often coloured green and grey, or green and brown
- wings have numerous tufts of erect scales

These moths usually have the labial palpi upturned; they are night-flying, coming readily to light.

The larvae feed on the leaves of fig trees, joining the leaves with silk or boring in the shoots or fruit. They pupate in an oval cocoon in the larval shelter with the pupal shell not protruding from the cocoon after emergence of the moth.

There are seven species in Australia, all found in tropical and subtropical rainforests, although one species attacks cultivated figs in suburbs. There are about 40 species worldwide.

Phycomorpha prasinochroa is found from Eungella, Queensland south to Nowra, New South Wales. The larvae bore in the tips, or tie young leaves together, of cultivated fig trees and also the sandpaper fig, *Ficus coronata* (Moraceae). Photo: John Stockard

Carposinidae

- small to medium-sized
- head smooth
- wings held back flat over body
- antennae held back along leading edge of wings at rest
- antennae simple
- forewings are narrower than the hindwings
- forewings often have tufts of erect scales

Most Carposinidae have palpi pointing forward in front of the head. They are usually grey but a few are green or black; they are night-flying, coming readily to light.

In Australia, the larvae of one species have been reared from the dying bark of damaged eucalypts or from the flowers of *Leucopogon* (Epacridaceae), and another has been reared from the woody fruit of *Hakea* (Proteaceae). One large species near Sydney feeds in the old flower spikes of *Banksia* (Proteaceae), and one pest species feeds on fruit of native trees and on citrus, guava and feijoa. The cocoon is in the larval shelter or in litter on the ground; the pupal shell is not protruded on the emergence of the moth.

There are 40 species recorded from Australia and 273 species worldwide.

Carposina smaragdias is uncommon but occurs on the Atherton Tableland south to near Townsville, Queensland. It has been found in rainforest vegetation fringing streams.

This species of *Sosineura* is found in the south-western corner of Western Australia. Nothing is known of its biology.

119

Epermeniidae

- small
- head rough-scaled
- wings held roof-wise at rest
- antennae held back along leading edge of wing at rest
- antennae simple
- both wings long and narrow with long scale fringes along trailing edges
- coloured drab brown or grey
- trailing edge of forewing with protruding tufts of dark scales

These moths have upcurved palpi and hind legs with stiff bristles. The adults are night-flying and come to light.

The larvae have been reared from the leaves of *Exocarpos cupressiformis* (Santalaceae) where they browse in the open. Others have been reared feeding in the fruit of quandong (*Santalum acuminatum*, Santalaceae) or within the green seed capsules of blackthorn (*Bursaria spinosa*, Pittosporaceae). The moths are widespread in Australia, usually in drier habitats.

There are 17 species recorded from Australia and about 70 species worldwide.

This is probably a species of the Epermeniidae although a few *Zelleria* in the Yponomeutidae look similar but generally sit head downwards. Species of Epermeniidae are found throughout Australia. The larvae feed on the seeds and fruit of trees and sometimes on the leaves.

Tineodidae

- small
- smooth head
- wings held out at right angles to body at rest and horizontally up off the substrate
- antennae usually simple, often longer than wings, held out at an angle from the head
- forewings with tip extended, pointed
- palpi sometimes long, projecting directly forward of head
- very long and fragile legs
- body raised well above substrate

Moths of this family are rarely encountered and most occur in the tropics and subtropics. The species *Cenoloba obliteralis* is unusual in having cleft wings, the tip of the forewing not extended and short antennae. The single species with pectinate antennae, *Epharpastis daedala*, is found only in the southern part of Western Australia as far east as Eucla.

The larvae feed between leaves of the foodplant joined with silk, and pupate in a cocoon spun in the same place. The pupal shell is not protruded with the emergence of the adult.

There are 11 species in Australia, with a further 18 species overseas, mostly in the Oriental region.

Tineodes adactylalis is widespread in southern Australia including Tasmania. Nothing is known of its biology. This is a male. Photo: Andreas Zwick

Cenoloba obliteralis is found from Innisfail, Queensland to Grafton, New South Wales. The larvae feed within the seeds of the mangrove *Avicennia marina* (Verbenaceae), while they are still attached to the tree. Unlike this species, most tineodids do not have incised wings but have long legs and sharply pointed forewings. This is a female.
Photo: Ian Common and Murray Upton

Alucitidae (Many-plume Moths)

- small
- smooth head
- wings held out at right angles to body at rest and horizontally up off the substrate
- antennae short, simple
- antennae held back under wings at rest
- long fragile legs
- moth has a dancing gait and raises and lowers wings a few times before coming to rest
- fore and hindwings deeply cleft into numerous narrow lobes

These moths are unmistakable in having their wings deeply cleft many times, in contrast to the Pterophoridae with a few deep clefts in each wing, and with *Cenoloba* in the Tineodidae with only one cleft in each wing.

The larvae form galls or feed in flower buds, shoots or fruit. Some pupate in the larval shelter; others leave the shelter and form a cocoon in litter on the ground. The pupal shell is not protruded from the cocoon when the moth emerges. The little

dance the moth does while raising and lowering the wings before it settles is characteristic.

Australian species have been reared from the plant families Bignoniaceae and Rubiaceae. They are found throughout the northern half of Australia in rainforests, monsoon forests, savannah and the arid zone. There are six recorded species in Australia (but certainly more unrecorded) and about 130 worldwide.

Alucita phricodes is found from the Atherton Tableland, Queensland to Batemans Bay, New South Wales. The larvae have been reared feeding on the buds and flowers of *Pandorea jasminoides* and *P. pandorana* (wonga vine) (Bignoniaceae).
Photo: John Stockard

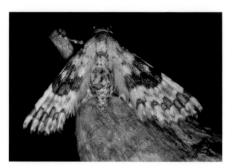

This medium-sized species of *Alucita* has been found in widely separated parts of Queensland and Western Australia. The larvae form galls on *Canthium* (Rubiaceae). Photo: Murray Upton

Pterophoridae (Plume Moths)

- small to medium-sized
- smooth head
- wings held out at right angles to body at rest and horizontally high up off the substrate
- antennae simple, fairly short
- antennae held out in front of head, horizontally and at a 90-degree angle to each other
- long, thin body
- very long and fragile legs
- wings narrow
- forewings deeply cleft once, hindwings deeply cleft two or three times

These are extremely fragile moths sometimes active in the day or at early dusk but some also come to light at night.

The larvae feed on flowers or occasionally on leaves openly during the day and have numerous small hairs on them. One species is common in gardens on geranium flowers and another, from the Snowy Mountains, feeds on *Gentianella* (Gentianaceae). *Platyptilia omissalis* is a brown moth whose larvae feed on the leaves of *Parahebe* (Scrophulariaceae). The pupae are elongate, without a cocoon, and rest fully exposed on the vegetation attached to a silk pad by hooks at the rear end. The adults have a weak, almost hovering flight.

There are about 30 recorded species in Australia and about 990 worldwide.

Xyroptila marmarias is found on the Atherton Tableland in Queensland but nothing is known of its biology. The larvae of most related species feed on flowers. The bright colours suggest that it is day-active. This is a female.

This fragile moth, *Cosmoclostis aglaodesma*, is found from the Atherton Tableland, Queensland south to near Taree, New South Wales. The deeply dissected wings are typical of the family. The moth has been reared from larvae feeding on the flowers of the white beech tree (*Gmelina leichhardtii*, Lamiaceae). Other pterophorids feed on flowers as larvae.

Hyblaeidae

- medium-sized to large
- head and body covered with smooth, short scales
- wings held flat back over body when at rest
- antennae fairly short, thread-like
- antennae held back over body at rest
- protruding palps
- wings broad
- very small head and eyes but stout thorax
- hindwing often contrastingly patterned in black and yellow, orange or pink

The hyblaeids have a well-developed unscaled proboscis but no hearing organs. They are most easily recognised by the very small head in relation to the very stout thorax and the protruding palps. The adults mostly fly at night and all the Australian species are found in tropical and subtropical rainforest.

The larvae cut and fold over a piece of leaf of the foodplant and silk it together as a shelter and when nearly fully grown they roll a complete leaf. The pupa is formed in the old larval shelter on the foodplant and is not protruded from the cocoon when the moth emerges.

Hyblaea puera is a pest that defoliates teak in South-East Asia and is also found in Australia where it feeds on other species of Verbenaceae.

There are five species recorded from Australia and 18 species in the Old World tropics.

Hyblaea ibidias is a medium-sized moth found from Eungella near Mackay, Queensland to Narooma, New South Wales. The hindwings, hidden here, are coral pink. The larvae have been reared from silken shelters on *Pandorea jasminoides* (Bignoniaceae), the bower vine. Photo: Ian Common

Thyrididae (Leaf Moths)

- medium-sized to large
- smooth head
- wings broad, held horizontally out at right angles to body, sometimes up off substrate, sometimes touching substrate
- forewings often warped towards tip
- antennae simple or slightly flattened
- antennae held back under wings when at rest
- wings often brown with a characteristic reticulate or net-like pattern
- body usually short and stout but slender in some small species

Thyridids are called leaf moths because the net-like pattern on the wings and their stance make them look like dead leaves. Most are brown but a few are red, yellow or white.

The larvae may form galls in stems of trees but some of the Australian species make shelters of silk on the green leaves of plants. The pupa may be found in the larval shelter or in leaf litter on the ground. Thyridids have been reared from a wide range of different plants.

In Australia most species occur in tropical and subtropical rainforest but with a few in eucalypt forest; none reach Tasmania. There are 56 recorded species in Australia and about 760 worldwide.

The medium-sized leaf moth *Canaea hyalospila* is found in rainforests from Cape York to near Mackay, Queensland. Nothing is known of its early stages. A very similar species is found in southern Queensland and northern New South Wales. Leaf moths are so named as they resemble leaves; this species has hyaline spots to let light through its wings, which helps to disguise it. This is a male.

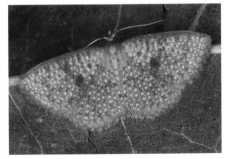

Aglaopus carycina is small to medium-sized, brightly coloured. It is found from the Atherton Tableland south to near Townsville, Queensland. Nothing is known of its biology. This is a female.

The large leaf moth *Oxycophina theorina* is found from Cape York south to near Brisbane, Queensland and also in New Guinea. Nothing is known of its early stages but it flies in rainforest. This is a female.

The small *Addaea subtesselata* is found from Brisbane, Queensland south to Taree, New South Wales in rainforests. The larvae join leaves of *Mallotus philippensis* (Euphorbiaceae) with silk and feed in the shelter so formed. Photo: John Stockard

This small, delicate little moth, a species of *Calindoea,* is found in rainforest from Heathlands to Iron Range, on Cape York Peninsula. They have long legs and rest on the green leaves during the day. This is a male.

Pyralidae

- small to large
- head usually smooth, sometimes rough-scaled
- wings held in many different poses, sometimes out at right angles to the body and horizontally, sometimes steep roof-wise, sometimes flattened roof-wise, sometimes rolled about body
- antennae usually simple but sometimes pectinate, usually about half length of wing, but sometimes very short or long
- antennae often held back over wings or body and together but sometimes along leading edge of forewing
- if wings held back at rest then hindwing often without intricate pattern and broader than forewing but if wings held out then the hindwing may be patterned and no broader than the forewing

This is one of the mega-diverse families of moths and the adults adopt many postures, behaviours and biologies. They are technically easily distinguishable from other moths by usually having a scaled base to the proboscis (found otherwise in the gelechioid families), abdominal hearing organs, and often visible maxillary palpi in addition to the prominent labial palpi that usually protrude forward directly in front of the head, although they are sometimes large and ascending. The legs may be long and fragile, or short. The pupal shell is extruded from the cocoon on emergence of the adult. There are many subfamilies and as many as possible have been illustrated. (The subfamilies Chrysauginae, Noordinae and Musotiminae have been omitted.)

The Pyraustinae, with mostly brightly coloured species, is the largest subfamily. Its

The small to medium-sized moth, *Heteromicta poeodes*, has been found in rainforest from just south of Cooktown, Queensland to the Allyn River near Newcastle, New South Wales. The group to which it belongs is usually associated with monocots such as palms, but nothing is known about this species. Subfamily Galleriinae.

Orthaga seminivea is a medium-sized moth found in rainforests from the northernmost parts of the Northern Territory, and northern Queensland south to Wollongong, New South Wales. The larval stage is unknown but, if it is like that of related species, it will silk together a bundle of leaves or twigs of the foodplant and live in the shelter so formed. Subfamily Epipaschiinae.

members (together with those of Pyralinae) have colour patterns on their hindwings. In other subfamilies, the moths have plain or two-tone hindwings.

Pyralid larvae have many varied habits. Typically they are concealed feeders, living in leaves tied with silk, in silken webs or in cases made of leaf fragments. They may also bore in trunks, stems, fruit or galls or live in a tunnel in leaf litter or the soil. Some subfamilies have a more specialised biology; the Crambinae and Schoenobiinae all feed as larvae on grasses or sedges while the Midilinae in Australia are all associated with the Capparaceae. The subfamily Nymphulinae is the only group of aquatic moths known and the larvae live underwater, feeding on water plants although the adults, at least in Australia, are all fully aerial.

There are many pest species: most of the meal moths are Pyralidae and others include the wax moths and pests of many crops such as the rice stem borers. The female rice stem borer moths have tufts of specialised scales at the end of the abdomen, which they cover the eggs with to help protect them. These scales, coming loose in flight, may cause irritation of the human throat if inhaled in sufficient quantity. The classical Australian exemplar for biological control is the pyralid, *Cactoblastis cactorum* (see p. 28), and its control of the weed prickly pear in southern Queensland continues to this day.

There are about 1100 recorded species in Australia but hundreds more known; there are about 16 000 species worldwide.

Styphlolepis agenor is found in the drier areas of southern Queensland and northern New South Wales. The larvae bore gregariously in the trunk of *Capparis mitchellii* (Capparaceae). This large moth is resting on a trunk, showing the extruded sawdust from other larval tunnels. Subfamily Midilinae.
Photo: Ted Edwards

Crocidolomia pavonana is a medium-sized moth, found from Africa and India to Norfolk Island. In Australia it is found in the Kimberleys, the top end of the Northern Territory and in eastern Australia from Cape York Peninsula south to Taree, New South Wales. The larva is known as the cabbage cluster grub and feeds on a range of plants related to cabbages and capers. This is a male. Subfamily Evergestinae.

Didymostoma aurotinctalis is a medium-sized, very delicate species found from Iron Range to the Cairns district, Queensland and in New Guinea. Nothing is known of its early stages. See-through wings occur frequently in the rainforest pyralids where the moths take on some of the colour of their backgrounds under the dim forest light conditions. This is a female. Subfamily Pyraustinae.

Pachynoa xanthochyta is a large, uncommon, but very pretty moth found in rainforests on Cape York Peninsula and extending as far south as Townsville, Queensland. Nothing is known of its biology. This is a female. Subfamily Pyraustinae.

Hygraula nitens is a small moth found very widely south from Brisbane, Queensland to Tasmania, and south of Broome in Western Australia. Remarkably for an aquatic moth it is found in arid as well as coastal areas and can tolerate mildly brackish water. It is also often seen at light a considerable distance from water. The larva has been reared from underwater cases made from leaves of *Potamogeton* (Potamogetonaceae) and *Zostera* (Zosteraceae). This is a male. Subfamily Nymphulinae.

This small moth, *Heliothela ophideresana,* is hiding a brilliant orange and black hindwing. It is found throughout Australia north of about the latitude of Canberra, Australian Capital Territory. It is found in rainforest and also in the most arid areas of central Australia. Nothing is known of its biology. This is a female. Subfamily Scopariinae.

This large, conspicuous moth, *Meroctena staintonii*, is found on Ambon and in New Guinea as well as in Queensland, from Cape York to near Townsville. It is found in rainforest but nothing is recorded of its biology. This is a female. Subfamily Pyraustinae.

This small moth, a species of *Mesolia*, is found in northern Queensland. It has an unusual resting posture. Related moths feed as larvae on grasses and sedges. Subfamily Crambinae.

This medium-sized moth, a species of *Curena*, is found in rainforest from the Atherton Tableland to near Townsville, Queensland. Nothing is known of its biology. This is a female. Subfamily Pyralinae.

This large, elegant moth, *Glyphodes stolalis*, is known from India to Fiji and in Australia in the Kimberleys, Western Australia, the top end of the Northern Territory and from Cape York, Queensland to Coffs Harbour, New South Wales. It has been found in rainforest and monsoon forest in Australia and overseas the larvae have been reared from several species of fig tree (Moraceae). This is a male. Subfamily Pyraustinae.

Margarosticha sphenotis is a small moth found from Cape York, Queensland south to Taree, New South Wales, and in the top end of the Northern Territory. It is an aquatic species but otherwise nothing is known of its biology. This is a male. Subfamily Nymphulinae.

Hyalobathra miniosalis is a medium-sized moth found from Sri Lanka through South-East Asia to Australia, where it is found in the top end of the Northern Territory and from the Torres Strait to Noosa, Queensland. It is a variable species and sometimes the red colour along the veins stands out strongly. It is found in rainforest and monsoon forest where the larvae have been found webbing leaves of *Glochidion lobocarpum* (Euphorbiaceae) with silk. Subfamily Pyraustinae.

131

This large, striking moth, *Pygospila tyres*, is found in the Kimberleys, Western Australia, the top end of the Northern Territory and in Queensland from the Torres Strait Islands south to Brisbane. Overseas it is found from India eastwards to Australia. Its biology is not reported from Australia but overseas the larvae feed on the plant family Apocynaceae. This is a female. Subfamily Pyraustinae.

Thesaurica argentifera is a glittering little moth found in rainforest from Cooktown to the Cairns area in Queensland. Nothing is known of its biology. Subfamily Odontiinae.

This large female, *Vitessa zemire*, is a striking moth found from the Molucca Islands eastwards to the New Hebrides and in Queensland from the islands of Torres Strait to Cardwell. Nothing is known of its biology but the similarity of related species suggests that it is part of a large mimicry group. Subfamily Pyralinae.

Conogethes punctiferalis is a medium-sized moth found from India and Japan eastwards to Australia, and from the Kimberleys, Western Australia, the top end of the Northern Territory and Cape York, Queensland to Kiama, New South Wales. The larvae are pests and are known to live in the flowers, fruits and shoots of a huge range of plant families. In Australia they have been reared from citrus, peach, macadamia, maize and sorghum among many other crops. This is a female. Subfamily Pyraustinae.

Herpetogramma licarsisalis is a medium-sized to large, extremely common pest species found from Europe, Africa, Asia and the Orient to the Pacific Islands. In Australia it is found in the top end of the Northern Territory and from Cape York Peninsula, Queensland to Batemans Bay, New South Wales. The caterpillars live in silk shelters in the ground and damage pasture and lawn grasses. This is a female. Subfamily Pyraustinae.

Hellula hydralis is a small, extremely common and widespread species in Australia found throughout the continent, including Tasmania. It is a vagrant to New Zealand but not established there. The larvae are a pest of brassicas, particularly the leaves, flowers and foliage of canola. A similar widespread species overseas, *Hellula undalis*, is found in the moist eastern coast of Australia south to Brisbane and Lord Howe Island. Subfamily Glaphyrinae.

This pretty little rainforest moth, *Aetholix flavibasalis*, is found from Cape York to Cardwell, Queensland. Overseas it is found from Sri Lanka to Australia. The larvae have not been reared in Australia but elsewhere they have been reared from a range of trees including mango, mangosteen and *Eugenia*. This is a male. Subfamily Pyraustinae.

Etiella behrii is a small moth found throughout Australia including Tasmania and overseas from China and Malaysia to the New Hebrides. It is a pest of legumes, the larvae living on the pods and seeds; it has been recorded from peanuts, soybeans, peas, lupins and lucerne seed. This is a male. Subfamily Phycitinae.

This medium-sized moth, *Hednota aurantiaca*, is found from Cape York south to near Maryborough, Queensland, but there are similar species found further south. The large genus *Hednota* is associated with grasslands and sedgelands as the larvae feed on these plants. Subfamily Crambinae.

The medium-sized *Tirathaba rufivena* is found throughout South-East Asia and eastwards to the Pacific Islands. In Australia it has been recorded from Cape York south to Rockhampton, Queensland. The larvae are pests feeding on the flowers of palms, including coconut palms. Subfamily Galleriinae.

Heosphora erasmia is a small to medium-sized moth found in open eucalypt country from Cooktown to Miles, Queensland. Its biology is unknown but the genus and related genera are known to have larvae that are borers in grasses. Subfamily Phycitinae.

Many species in the subfamily Schoenobiinae look similar. They are called rice stem borers and cause significant damage to rice crops in Africa, Asia and northern Australia.

This large moth, *Terastia subjectalis*, is found from India eastwards to Australia where it is found from Broome, Western Australia through the coast and arid zones of northern Australia south to Brisbane, Queensland and Lord Howe Island, New South Wales. The larvae bore in the branches, shoots and pods of *Erythrina* (coral tree, Fabaceae). When dead, the adults have a strong spicy smell. This is a male. Subfamily Pyraustinae.

Cardamyla carinentalis is a large moth found from the Torres Strait Islands, Queensland to Kiama, New South Wales. The larvae have been found feeding on *Pennantia cunninghamii* (Icacinaceae) but A.W. Scott and his daughters in the 1850s illustrated it feeding on *Cassine australis* (Celastraceae). This is a female. Subfamily Pyralinae.

Syntonarcha iriastis is a small to medium-sized moth found from the Pilbara, Western Australia through northern Australia south to Kiama, New South Wales. The larvae have been reared from the flowers of several species of *Melaleuca*. This is a female. Unusually for a moth, males of the species call in ultrasound to attract females using a scraper and file sited near the tip of the abdomen. Subfamily Odontiinae.

Niphopyralis chionesis is a small moth found in the Kimberleys, Western Australia, the top end of the Northern Territory and in Queensland from Cape York to Yeppoon. This genus extends from Australia to Sri Lanka and India. An old record of their biology from Java indicates a close association with green tree ants but the nature of this association is not clear. In the adult the abdomen is held erect in a curious manner. Subfamily Wurthiinae.

Aquatic moths

Most people regard moths as purely land insects but members of the pyralid subfamily Nymphulinae are aquatic, that is, the larvae are aquatic although the adults behave like any other moths.

Most of the 50 Australian species live in the coastal parts of New South Wales and Queensland and the adults are found close to streams. The species *Hygraula nitens*, however, is widely distributed, even in inland Australia, and can be found at great distances from streams. It has also been found living in brackish water in coastal lakes.

Little is known about Australian species, but those in which the biology is known construct cases out of severed leaf pieces of aquatic plants, or build shelters of reed stems. The larvae obtain oxygen by four methods. Very small larvae may obtain adequate oxygen by simple diffusion through the skin. Larger larvae may capture oxygen from the surface on hairs or roughness of their skin, and store it underwater. Others have slender unbranched filamentous gills in two rows along each side of the body. Most commonly in Australia, the larvae have branched tracheal gills arising from nearly every segment of the body.

Overseas, two feeding strategies are known; larvae either live in shelters while feeding on the leaves of aquatic plants, or they live under a silk tent and graze on algae. Several species of adult nymphulines have been observed clustering on leaves or twigs overhanging water, but why they do this is not known. The potential of these moths to act as indicators of water quality has often been overlooked but, like some other small and well-known aquatic groups, they can play a significant role as important indicator species.

This small aquatic moth, *Parapoynx diminutalis*, (Pyralidae), is found from the Pilbara, Western Australia through northern Australia and south to Myall Lakes, New South Wales. Overseas it occurs as far west as Sri Lanka and India and north into China. Many of the aquatic moths are very delicate. This is a female.

This is the larva of the aquatic moth *Parapoynx diminutalis*. It is covered with a mass of tracheal gills. Many aquatic larvae make shelters from the water plants they feed on. Overseas it has been reported from several water plants, including rice.

Geometridae (Loopers, Inchworms)

- small to large
- smooth or rough-scaled head
- wings usually outstretched from body pressed flat against the substrate, but sometimes back along body roof-wise or rolled about body, and a few with wings held upright together like butterflies
- antennae in male usually pectinate, sometimes pectinate in female but usually simple
- antennae held back under forewing when at rest
- wings very broad and only rarely with forewings narrower than hindwings
- usually with intricate patterns on both fore and hindwing, rarely without pattern on hindwing and then hindwings covered at rest
- wing pattern usually consists of a large number of wavy more or less parallel lines crossing the wing and without prominent circles

This is one of the mega-diverse families but its members have fewer different habits and behaviours than those of Pyralidae. They are most easily recognised by having abdominal (rather than thoracic) hearing organs, an unscaled proboscis and a small pincushion-like organ (chaetosema) above the eye.

There are five main subfamilies: Ennominae, Geometrinae, Oenochrominae, Larentiinae and Sterrhinae. Not illustrated in the following pages is a small subfamily, Archearinae, which may not really be present in Australia although a few alpine Tasmanian species have been included in it. Also not illustrated is another small subfamily found in the tropical rainforests, Desmobathrinae.

The subfamily Ennominae has a very

The larva of *Aeolochroma metarhodata*, a typical looper or inchworm larva, moves by drawing the rear to the head and then advancing the head. The reduced number of prolegs allows the body to bend upwards a long way, which means more ground is covered when walking. Subfamily Geometrinae. Photo: Bob Jessop

The distinctive larva of *Dysphania numana* is more bulky than most loopers but retains the looper features. It feeds on *Carallia brachiata* (Rhizophoraceae) and is found in rainforest in northern Queensland and the Northern Territory. Subfamily Geometrinae.

Milionia queenslandica is a large, spectacular moth that belongs to a New Guinean group of brilliant day-flying moths with larvae that feed on conifers. The Australian species are confined to rainforest and have not been reared. This one is found from Cooktown to Mackay, Queensland. The Australian species come to light at night and do not seem as strongly day-flying as those in New Guinea. This is a male. Subfamily Ennominae.

Sauris malaca is a medium-sized to large moth confined to rainforest on the Atherton Tableland, Queensland and south to Kiama, New South Wales. The hindwings of the male (not visible here) are grey and narrowed, folded and with blisters that help to distinguish it from a suite of similar species. The larvae have been reared from litchi and red cedar leaves. This is a male. Subfamily Larentiinae.

Eucyclodes pieroides is a medium-sized, common moth found across the Northern Territory from the Kimberleys and from Cape York, Queensland to Wingham in central New South Wales. The larvae feed on a very wide variety of plants and have a series of grotesque flanges on their sides. This is a female. Subfamily Geometrinae.

The medium-sized *Protuliocnemis biplagiata* is very widespread from Sri Lanka to New Guinea, in Queensland as far south as Tully and in New Caledonia. Overseas, the larvae have been reared on *Acacia* and in Australia it is found in rainforest. This is a male. Subfamily Geometrinae.

large suite of grey species in southern Australia with many endemic genera that rest with wings roof-wise back over the body and are without pattern on the hindwings. The Larentiinae, as in other parts of the world, are most diverse in alpine habitats although there is a large group of small species in tropical and subtropical rainforests.

Most of the geometrids are night-flying but many of the Larentiinae are day-active. The common names for the family refer to the larvae, the well-known looper caterpillars. Using their true feet at the front and a reduced number of false feet at the back, they draw the back up to the front, thus looping the body, and then extend the front as an effective means of walking. They all

GEOMETRIDAE

Oenochroma quardrigramma belongs to a group of stout-bodied spectacular moths found widely in Australia but barely represented overseas. This large species is found in rainforests from Mossman, Queensland to Batemans Bay in New South Wales. Nothing is recorded of its biology but the larvae of many relatives feed on Proteaceae. This is a male. Subfamily Oenochrominae.

Abraxas expectata, like most moths in the genus, is large and boldly coloured in black and white. Some species often fly in rainforest during the daytime. Their contrasting colours help protect them and they may also contain poisons. *Abraxas expectata* is found from Cooktown to Cairns and on the Atherton Tableland, Queensland. This is a male. Subfamily Ennominae.

Crypsiphona ocultaria is one of the most common moths in Australia. It is found throughout Australia (with the possible exception of central Australia) and is often seen resting on walls of suburban houses. The undersides of the wings of this large moth are white, boldly marked in black and crimson. The larva is green and feeds on the leaves of eucalypts and rests exposed, twig-like, in the foliage. This is a male. Subfamily Geometrinae.

Aeolochroma turneri is a very common moth found in the Darwin area and from Cape York south to Bundaberg, Queensland. It is large and stout compared to the very fragile emeralds that are closely related. It seems unusually well adapted to resting on lichen-covered branches. This is a male. Subfamily Geometrinae.

feed openly on leaves, often adopting twig-like postures and many species have become wonderful mimics of twigs, stems, leaves and bark. The larvae pupate in a sheltered position, in litter or in the soil in a cocoon from which the pupal shell is not extruded on the emergence of the moth.

Geometrids are specialised aerial feeders and feed on a vast range of green plants. There are relatively few pest species but some occur in large enough numbers to be pests of crops and orchards and timber plantations.

There are 1300 recorded species in Australia with hundreds more known and more than 21 000 worldwide.

Alloeopage cinerea is a large moth found in rainforest from Cape York south to the Atherton Tableland, Queensland. It is also found in New Guinea. It rests during the day on leaves but nothing is known of its biology. This is a male. Subfamily Geometrinae.

This species of *Neogyne* has a very unusual resting posture for a geometrid, with its wings rolled and not pressed closely to the surface. This medium-sized moth is found in the monsoon forests on Cape York Peninsula and west of the Atherton Tableland, Queensland. Nothing is known of its biology. This is a male. Subfamily Ennominae.

Chrysolarentia insulsata is a medium-sized, common moth found in eucalypt forests and open ground on the tablelands of New South Wales, and in Victoria, Tasmania and the south-eastern part of South Australia. Nothing is known of its biology. Subfamily Larentiinae.

This large, rainforest moth, *Eumelea rosalia*, is found from Cape York to Mackay, Queensland. It also occurs widely in the Oriental region. Nothing is known of the larva in Australia but overseas it feeds on the euphorb, *Mallotus,* and on ginger. One has been reared from a pupa on a palm at Townsville. Subfamily Oenochrominae.

This large moth, a species of *Paraterpna*, is found in wet forests from the mountains near Brisbane, Queensland to the Grampian Mountains in western Victoria. Larvae have been reported feeding on *Leptospermum* leaves. This is a male. Subfamily Geometrinae.

Xenomusa metallica is found in rainforests on the Atherton Tableland, Queensland and south to Kiama, New South Wales. The curved leading edge of the wing, the drawn out tips of the forewings, and the wing markings all help to camouflage the moth in dead leaves. Nothing is known of its biology. Subfamily Ennominae.

This small moth, a species of *Chrysocraspeda*, is found only in the rainforests of the Atherton Tableland, Queensland. This is a male. Subfamily Sterrhinae.

Physetostege miranda is a large moth found only around Cairns, Queensland and in Papua New Guinea. Its biology is unknown. This is a male. Subfamily Oenochrominae.

Antitrygodes parvimacula is a medium-sized to large, spectacular moth found in rainforests in New Guinea and in Queensland from Cape York Peninsula south to Babinda. Related species overseas have caterpillars that feed mostly on plants of the family Rubiaceae. This is a male. Subfamily Sterrhinae.

This medium-sized rainforest moth, *Corymica pryeri*, is found widely in the Oriental region and in Australia from Cape York to Paluma, Queensland. Nothing is known of its biology in Australia but overseas it has been reared on several genera of plants in the Lauraceae. This is a male. Subfamily Ennominae.

Oenochroma vinaria is a very common moth, often found in native gardens, from the Atherton Tableland, Queensland south to Tasmania, South Australia and southern Western Australia. The larvae of this large moth are easily found on many genera of Proteaceae including *Grevillea* and *Hakea*. This is a male. Subfamily Oenochrominae. Photo: John Stockard

Poecilasthena sthenommata is a medium-sized, very fragile moth that rests on leaves in the daytime. It is found in rainforest from Atherton south nearly to Townsville, Queensland. This is a female. The early stages are unknown. Subfamily Larentiinae.

Pingasa blanda is a large moth found in New Guinea, in the Northern Territory and in Queensland from Cape York south to Mackay. The peculiar markings on the wings near the base certainly look as though the moth is mimicking something with prominent legs, perhaps a spider. This is a female. Subfamily Geometrinae.

Dichromodes personalis is a medium-sized moth found in Western Australia from Jurien Bay to Esperance. It is very common but nothing is known of its early stages. Other species in this very large Australian genus feed on Myrtaceae. This is a female. Subfamily Oenochrominae.

Metallochlora lineata is a medium-sized, common moth known from New Guinea and in northern Queensland from Cape York Peninsula south to near Rockhampton. The larva and foodplant are unknown. The dusting of silvery scales is characteristic of this species. This is a male. Subfamily Geometrinae.

Niceteria macrocosma is a large moth found in wet forests from the Atherton Tableland, Queensland to Adelaide and Tasmania. This male has been disturbed from its resting posture to show the bright lemon and black hindwings. The larva rests on the eucalypt foodplant, feeding on the leaves, and is green with pale green or cream lateral and mid-dorsal lines. Subfamily Ennominae.

Bracca rotundata is a large moth found in rainforest from Cape York Peninsula to Eungella, west of Mackay, Queensland. A very similar moth is common in rainforest in south-eastern Queensland and north-eastern New South Wales. Subfamily Ennominae.

Dithalama cosmospila is a medium-sized, delicate moth found from central Queensland to southern New South Wales. Nothing is known of its biology. The pose with flattened wings is typical of a geometrid. This is a male. Subfamily Sterrhinae.

Cernia amyclaria, a medium-sized moth, is widespread from the Pilbara, Western Australia through the Kimberleys and top end of the Northern Territory and in Queensland south from Cape York Peninsula to Batemans Bay, New South Wales. It can be common but nothing is known of its biology. Subfamily Oenochrominae.

Oenochroma polyspila is a medium-sized and rather variable moth in the extent of the spotting on its wings; some individuals have the pink suffused with yellow. It is found in tropical eucalypt woodland from Cape York Peninsula to Townsville, Queensland. The larvae of related species feed on Proteaceae. This is a male. Subfamily Oenochrominae.

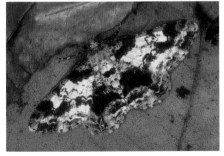

Cleora illustraria is found in the Solomon Islands and New Guinea and in Australia from Cooktown, Queensland south to eastern Victoria. This large moth is extremely variable in colour and pattern. The larvae have been reared on *Leptospermum*. This is a male. Subfamily Ennominae.

This medium-sized moth is a species of *Pachyplocia*, a genus that occurs very widely in northern Australia south to Alice Springs and central New South Wales. Nothing is known of its biology but it does seem to occur consistently where *Melaleuca* is growing. This is a male. Subfamily Ennominae.

145

Oenochroma cerasiplaga is a medium-sized moth found in Western Australia from Fitzgerald River east to Israelite Bay. Nothing is known of its biology but larvae of related species feed on Proteaceae. Subfamily Oenochrominae.

This striking species, *Arcina fulgorigera*, is a medium-sized moth found in eucalypt forests in south-western Western Australia from Perth to Albany. Nothing is known of its biology. This is a female. Subfamily Oenochrominae.

Capusa stenophara is one of a large group of long-winged geometrids that fold their wings at rest while most geometrids rest with wings flat against the substrate. This large moth is known from central and southern New South Wales and the Australian Capital Territory. The larvae of a very similar species feed on eucalypts and are green with prominent dark brown spots. Subfamily Ennominae.

This large moth, a species of *Hypographa*, is found in Western Australia from Perth to Mt Ragged. It is possibly the same species as a very similar looking moth found from Kiama, New South Wales to Kangaroo Island, South Australia in the east. Larvae of the eastern species have been reared from *Hakea* (Proteaceae). This is a male. Subfamily Oenochrominae.

Drepanidae

- medium-sized to large
- smooth head
- wings held outstretched close to substrate at rest
- antennae held beneath wings at rest
- antenna pectinate and short, or simple and long
- body short, stout if antennae are pectinate
- tip of forewing extended to a broad curved tip if antennae pectinate, or a small tip if antennae simple

This is a small family with two very different groups in Australia. The *Hypsidia* group contains large moths with simple antennae and barely extended tips to the forewing. There are two very brightly coloured species in the rainforests of northern Queensland and four plain species in the south-west of Western Australia. Nothing is known of the biology of this group. The second group contains the typical 'hook-tips' found in other parts of the world with stout bodies and broadly extended curved tips to the forewing. These are found in rainforest as far south as Sydney.

Australian drepanid species have not been reared but overseas the larvae rest

The distribution of this large, brilliant moth, *Hypsidia erythropsalis*, is limited to rainforests from Cape Tribulation to Babinda and parts of the Atherton Tableland, Queensland. Nothing is known of its biology.

Hypsidia robinsoni is a large moth found in the higher parts of the Atherton Tableland, Queensland. Nothing is known of its biology but the adults have been collected in rainforest. Photo: Ian Common

openly on the vegetation and the rear of the larva is extended, protrudes, and is raised while the larva rests. The moths are all night-flying and have hearing organs at the base of the abdomen like the geometrids but of a different structure.

Drepanidae is a small family with 10 species in Australia and about 650 worldwide.

Astatochroa fuscimargo is a small moth found in rainforest from Cape York to Brisbane, Queensland. Nothing is known of its biology. This is one of the 'hook-tipped' moths belonging to the subfamily Drepaninae, more familiar to people outside Australia.

Uraniidae

- large to very large
- smooth or rough-scaled head
- rests with wings outstretched flat against substrate or sometimes rolled and twig-like
- antennae simple or flattened
- antennae held back under wings at rest
- wings broad
- larger species white with hindwing with a sharp angle in middle of the margin, or coloured with a protruding tail; medium-sized species with hindwing usually rounded, brown or grey

The family Uraniidae contains two very different groups: the usually large Uraniinae and the usually medium-sized Epipleminae. Both subfamilies are confined to tropical and subtropical rainforest in Australia except for *Phazaca interrupta*, which occurs widely in the arid zone.

The Uraniinae may be white with a faint net of brown lines, green and pink like *Alcides*, or brown and white like *Lyssa*. Most are day-flying although *Lyssa* flies at night. The Epipleminae are smaller, drab coloured, including a few that rest in an irregular manner, either with wings splayed and separated with forewing or hindwing rolled.

Moths of the Uraniidae family, like the Geometridae and Drepanidae, have abdominal hearing organs but they differ in detailed structure and are day-flying. The larvae feed exposed in the leaves of the foodplant and the pupae are found in litter on the ground or in crevices.

There are 36 species in Australia and about 700 worldwide.

Phazaca decorata is found only in Australia where it is known from the top end of the Northern Territory and from Cape York, Queensland south to Lismore, New South Wales. A small moth, it occurs in rainforest and wet fringing forest along streams. The wings are rolled to look like a dead twig—a characteristic of the subfamily Epipleminae which, unlike many other uraniids, fly at night and rest in the day. The early stages are unknown.

Phazaca interrupta is a species of arid and semi-arid Australia. This small to medium-sized species is found in the Pilbara, Western Australia eastwards through the centre to Coen and inland Queensland to the Darling Downs and Broken Hill to Queanbeyan, New South Wales. Sometimes it is found in moister areas as at Cooktown or Brisbane, Queensland. The larvae have been found on *Canthium oleifolium* (Rubiaceae). This is a male.

149

Alcides metaurus is a very large and conspicuous day-flying moth, most readily seen in the early morning when it visits flowers to feed. It is found in rainforests from Cape York south to Mackay in Queensland. The larva feeds on *Omphalea queenslandiae* and several species of *Endospermum*, all of which are euphorbiaceous vines or trees.

Lyssa macleayi is a very large moth found in New Guinea and Indonesia west to Tanimbar Island. In Queensland it occurs in rainforest from Cape York to Tully. The larvae have been found on *Endospermum medullosum* (Euphorbiaceae), and *E. myrmecophilum* is probably also a foodplant. Unlike *Alcides*, these moths are night-flying. This is a male.

THE BOMBYCOID FAMILIES

The bombycoid families include all the families from Lasiocampidae to Sphingidae (pp. 152–170).

The Sphingidae, hawk moths, are well known with their very strong, rapid flight, their long proboscis, their streamlined bodies and their ability to hover before flowers. They have no hearing organs.

The remaining bombycoid families are very different; they are hairy moths, also with no hearing organs. They have specialised in sedentary behaviour, are mostly unable to feed or drink, and are short-lived. They lay large eggs almost anywhere and all have broadly pectinate antennae in the male with the pectinations usually long until the tip of the antenna. The females emerge from the cocoon with almost all their eggs ready to lay and are often much larger than the males. The males can detect the scent of the female with great efficiency—females may call males from up to a kilometre away. They have also specialised in strong extensive cocoons of silk; almost all the moths whose silk is used by humans come from this group.

Lasiocampidae

- large
- 'furry' or 'woolly' moths with head and body well covered in long hair-like scales
- wings held roof-wise back over body when at rest
- antennae in male broadly pectinate, often gradually bent in the middle
- pectinations 'droop' downwards
- antennae held back under wings at rest
- wings broad
- the hindwings are usually much smaller than the forewings and usually without so much colour pattern
- females usually with very large bodies in which almost all eggs are ready to lay when the moth emerges from the cocoon

The moths in this family have no hearing organs, and the proboscis is reduced or absent. Lasiocampids may be recognised by the presence of a pincushion-like organ on the underside of the palpi (when viewed with a strong lens), which is found in no other family. They also lack the hook-and-bristle mechanism between forewings and hindwings.

Lasiocampids are mostly night-flying, but a few species have day-active males and night-active females. In these the females 'call' the males during the day and the males probably rarely fly except when detecting the female scent at which time they fly with great rapidity. Some species with day-active males have reduced eyes, a feature often found in moths where the eyes remain functionally 'night-time eyes' but the moths now fly in the day. The female also has reduced eyes, even though they fly at night to lay eggs.

The larvae are very hairy, often flattened, and often with spreading hairs on the sides of the larvae extending down to touch the branch they they rest on, which helps to blur the outline of the larva. Some rest during the day on twigs and branches while others hide beneath bark and come out to feed at night.

The larvae of *Entometa* have the membrane between the segments of their thorax coloured in black and blue, and when disturbed they rear up exposing the coloured skin. They pupate in strong cocoons of tough silk, in the foliage or in sheltered places in litter or under bark. Some *Entometa* have oval, rounded cocoons (sometimes coloured bright green) hanging from a twig by a thick silken rope. The pupa does not protrude from the cocoon on emergence of the moth. Overseas, some larvae cause a skin rash if they are handled but no Australian species is reported to do this.

There are about 70 species in Australia and about 1500 species worldwide, particularly in Africa.

This widespread, medium-sized moth is a species of *Pararguda*. It is found in tropical eucalypt forest from the Kimberleys to the top end of the Northern Territory, and in Queensland from Cape York Peninsula south to near Mt Garnet. Nothing is known of its larvae or foodplant. This is a female.

This small moth is a species of *Genduara* belonging to a suite of species represented by *G. subnotata*. Members of this group are found in most parts of mainland Australia but are particularly plentiful in the arid centre. It is possible that the larvae feed on mistletoes. This species comes to light but some related ones are day-flying. This is a male.

This large moth, *Entometa fervens*, known from New South Wales, Victoria and South Australia, is easily confused with several related species. The larvae, which feed on eucalypts, are long, grey and covered with short fine hairs with longer hairs at the side to make the flattened larva look like part of the branch-let upon which it rests. The white cocoon is formed in a cluster of leaves of the foodplant silked together. This is a male.

In the genus *Pinara* the males and females are very different in size, shape and colour. The caterpillars feed on eucalypts. This medium-sized to large species is found in south-western Australia but others occur in the moist coastal areas throughout Australia. Females rarely come to light but they move about at night laying eggs. Males almost never come to light and fly extremely fast during the day. Rearing them from eggs is the best way to find which males and females belong to the same species. This is a female.

Anthelidae

- large
- 'furry' or 'woolly' moths with head and body well covered in long hair-like scales
- wings held outstretched or back roof-wise over body at rest
- antennae in male broadly pectinate to the tip, never bent
- antennal pectinations 'droop' downwards
- antenna held back under wings at rest
- wings broad
- hindwings usually nearly as large as forewings and usually with a pattern of colours
- females usually with very large bodies in which almost all eggs are ready to lay when the moth emerges from the cocoon

Anthelids have no hearing organs, and the proboscis is reduced or absent but they have a hook-and-bristle mechanism on the forewings and hindwings. They are night-flying, except for *Anthela connexa* males that fly very rapidly in the sunshine in Tasmania.

The moths often fly during the colder parts of the night, many coming to light after midnight. One Tasmanian species of *Pterolocera* appears in traps between midnight and dawn in cold rainy conditions in April in the far south-west of the state.

Anthela excellens is found in wet forests from northern Queensland to southern New South Wales. These large moths are very variable in colour ranging from grey-green to yellow to brick red but may be distinguished by the contrasting grey head. The very hairy larvae feed on leaves of various *Acacia* species. This is a male.

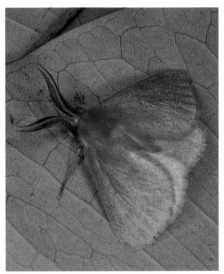

The large moth *Anthela ocellata* is very common from Brisbane, Queensland along the coast and tablelands to Adelaide, South Australia, and in Tasmania. The hairy larva feeds on a wide range of native and introduced grasses. This is a male. Photo: Bob Jessop

This medium-size to large species of *Anthela* is found in northern Queensland. Nothing is known of its biology. There are several very similar brick-red species and they are difficult to distinguish from one another. This is a male.

The adults of some species can be very variable in colour and this has caused many problems in identifying them in the past. The larvae are very hairy; in some the hairs may appear short and matted in patches but they are usually long and fine. Viewed from the front, the larvae usually have a vertical stripe of lighter colour on their head. The larvae feed exposed at night and some may seek shelter during the day. They pupate in a sheltered place, under bark, beneath logs, in the forest litter or underground. The cocoon is usually strong but not as tough as most lasiocampids. Many anthelid species feed on grasses, or on acacias or eucalypts.

The large anthelid *Chelepteryx collesi* has huge larvae banded in black, white and brick red. The larvae are armed with numerous tufts of short and stout but sharp hairs. They feed on eucalypts over summer and seek a sheltered place to spin their large whitish bag-like cocoon in early autumn. They often choose the undersides of over-arching branches, eaves or sometimes post boxes. After spinning its cocoon the larva slowly wriggles around until it has broken off all the hairs, which it then works through the cocoon until the cocoon is peppered with these sharp vicious spines. Anyone handling these accidentally ends up pierced with spines that are very uncomfortable until they are all individually pulled out. Usually victims suffer no or little allergic reaction but there are some records from Canberra of the hairs producing anaphylactic shock.

The family Anthelidae is confined to Australia and New Guinea. There are 74 species in Australia and about 12 in New Guinea with many species still unrecorded.

This moth belongs to a complex of species called the *Anthela acuta* group, which can be very variable in both colour and pattern. This species is from the rainforests of northern Queensland and its larva and foodplant are unknown. This is a male.

The larva of *Anthela varia* feeds on eucalypts and is very commonly found resting on the trunk of a tree or wandering away from the tree to find a sheltered place to spin its cocoon. The large pale grey or white cocoons of loose silk are often found under bark. Photo: Bob Jessop

This very large moth, *Chelepteryx collesi*, is found from central Queensland south to the wetter areas of Victoria. The very large banded and spiny caterpillars feed on eucalypts and spin large, loose cocoons, with spines protruding, which are often found in letter-boxes or under eaves. This is a male. Photo: Bob Jessop

Anthela postica is a large moth found from Sydney, New South Wales, to Gippsland, Victoria, but is rarely seen. The larvae have been reared on *Acacia*. This is a male. Photo: Bob Jessop

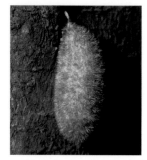

The cocoon of *Anthela ariprepes* shows the larval hairs extruded through the silken wall of the cocoon for added protection. This moth is found in arid southern Queensland and New South Wales. The larvae feed on *Acacia*. Photo: Bob Jessop

This large female *Anthela nicothoe* occurs in the cool forests of south-eastern Australia as far north as Stanthorpe, Queensland. The large hairy caterpillars feed on *Acacia* and the cocoons, bristling with sharp spines, are found under loose bark. Photo: Bob Jessop

Eupterotidae

- medium-sized to large
- 'furry' or 'woolly' moths with head and body well covered in long hair-like scales
- wings held partly out but low roof-wise or flattened over body at rest
- antennae in male broadly pectinate to the tip, never bent
- antennal pectinations 'droop' downwards
- antennae held back under wings at rest
- wings broad
- hindwings usually nearly as large as forewings and usually with a pattern of colours
- except in *Cotana*, the tip of the forewing is extended
- except in *Cotana*, the folded part of the hindwing, next to the body, has an enhanced colour pattern
- females usually with very large bodies in which almost all eggs are ready to lay when the moth emerges from the cocoon

Cotana neurina is a large species found from Cape York to Iron Range on Cape York Peninsula, Queensland but its biology is unknown. The larva of *C. serranotata* from the Cairns area has been reared from *Melaleuca quinquenervia* (Myrtaceae). The females of this genus look very different from the males and are much larger. This is a male.

Panacela lewinae is a medium-sized moth found in eucalypt forests from near Brisbane, Queensland to Bega, New South Wales. The larvae feed on *Eucalyptus, Angophora, Lophostemon* and *Syncarpia* (Myrtaceae) and live in silken bags in the leaves of the tree. The hairs of the larvae can cause severe skin irritation if handled. This is a male. Photo: Ian Common

This is a small family with three very different groups in Australia. *Ebbepterote* is a very large moth found only in tropical rainforest and its biology has not been recorded. The *Cotana* group is known from rainforest margins and tropical woodland in northern Queensland; males and females are very different. The *Panacela* group contains medium-sized moths, brown in colour, which are found in paperbark and eucalypt forests in eastern Queensland and New South Wales.

In the *Panacela* group the larvae are hairy with a short, round tuft on the back of each body segment. Viewed from the front, the head has a round yellow or white patch in the middle, in contrast to the vertical line in Anthelidae. They live gregariously in clusters of leaves silked together in a eucalypt or in a silk nest at the base of the tree; some may live in clusters openly on leaves. The larval hairs can cause skin rashes in humans if handled carelessly. Pupation is in a silken cocoon in the leaf litter.

This family is well represented in Africa but the subfamily Panacelinae is Australian. There are eight species recorded from Australia and about 400 worldwide.

Bombycidae

- medium-sized to large
- 'furry' or 'woolly' moths with head and body well covered in long hair-like scales
- wings held partly out but flattened at rest
- antennae in male very short, broadly pectinate to the tip, never bent
- antennal pectinations 'droop' downwards
- antennae held back under wings at rest
- wings broad
- hindwings not nearly as large as forewings and usually with little pattern of colours
- the tip of the forewing is extended in the commercial silkworm moth but not in native species
- the folded part of the hindwing next to the body has an enhanced colour pattern
- females usually with very large bodies in which almost all eggs are ready to lay when the moth emerges from the cocoon

These moths are distinguished from the Eupterotidae by a feature of the veins in the forewings and the presence of a horn on the larva. This family contains the silkworm, source of almost all the silk of commerce. The silkworm, *Bombyx mori*, was domesticated in China several millennia ago and does not exist in the wild where its closest relative is *Bombyx mandarina*.

In Australia the Bombycidae are represented by two species, one in the rainforests of Cape York Peninsula and the other in rainforest in southern Queensland and northern New South Wales. The larvae have several fleshy filaments on the body and a long erect horn near the rear end. They are not obviously hairy although they do have very short hairs. The larvae of the New South Wales species feed openly on fig tree leaves. (Figs are in the same family as mulberries, the silkworm foodplant.) They spin a neat green cocoon on a leaf.

There are about 350 species worldwide.

Bombyx mori is the commercial silkworm, although only used in schools in Australia. Originally from China, this moth is so domesticated that the species is not known in the wild. The silkworms feed on mulberry leaves, *Morus* spp. (Moraceae). This is a female. Photo: Peter Marriott

Carthaeidae

This family has but a single species in the world. *Carthaea saturnioides* is a large moth, which is easy to recognise because of its very distinctive colours and pattern. It is a typical 'wooly' bombycoid but has a strong hook-and-bristle mechanism on the forewings and hindwings. It also has a strong proboscis and the young larva has an erect horn near the rear end. The adults fly at night, sometimes very late.

Carthaea saturnioides is found in south-western Western Australia where the larvae feed on *Banksia* and *Dryandra* (Proteaceae).

The large *Carthaea saturnioides* is found in Western Australia from Geraldton south and east to the Cape Arid National Park. The larva feeds on *Dryandra* and *Banksia*. The adults fly at night in November and may be locally common. Photo: Andreas Zwick

Saturniidae (Emperor Moths)

- very large
- 'furry' or 'woolly' moths with head and body well covered in long hair-like scales
- wings held partly out but flattened at rest
- antennae in male very short, broadly pectinate to the tip, never bent
- antennal pectinations stiffly projecting from each side of the antenna in one plane
- antennae held partly up and out and back parallel to the front of the forewing
- both wings have either a triangular window or an eye-spot in the middle but in some the hindwing may just have a dot
- tip of forewing often broadly extended
- hindwings smaller than forewings but well patterned
- the body is very small for the size of the moth

This is a charismatic family with a huge band of devotees who rear emperor moths all over the world. They are often called silk moths but commercial silk comes from a bombycid, although there are several wild silks produced by saturniids.

The emperor gum moth and the Hercules moth are the best known Australian species. The moths may be distinguished by their distinctive antennae and the frequent development of eye-spots on the wings. They lack the hook-and-bristle mechanism on forewings and hindwings and the proboscis is minute or absent.

The large larvae are colourful and usually have sparse, small hairs and fleshy, raised turrets with longer hairs. They feed exposed on the vegetation during the day, living singly and pupating singly or in clusters when they pupate in a mass of cocoons. The cocoons are usually on the trunk of a tree, in the foliage, in crevices or under bark, and in one case in the litter on the ground. The thick, hard, oval cocoons of *Opodiphthera* are made of tough silk while

the atlas moth, *Attacus dohertyi*, which is found near Darwin, and the Hercules moth, *Coscinocera hercules*, have larger, looser, softer cocoons spun among leaves.

Only 15 species have been reported from Australia with about 1500 worldwide.

This very large Hercules moth, *Coscinocera hercules*, occurs in rainforest in New Guinea and in Australia, from Cape York south to Ingham, Queensland. It may fly at all times of the year but mostly during the wet season. This is a male.

161

Opodiphthera eucalypti is found in the top end of the Northern Territory, in Queensland from the Atherton Tableland to near Townsville, and from southern Queensland to Melbourne, Victoria. It is very variable in colour from grey, through straw-coloured, to almost brick red. Many people are familiar with its hard silvery grey cocoons on the trunks of the foodplants: eucalypts and the introduced peppercorn tree. This is a male.

Syntherata janetta is found in the Kimberleys of Western Australia, the top end of the Northern Territory and from Cape York, Queensland to Sydney, New South Wales. The colour and patterns of this very large moth are variable. It may be yellow, orange, grey, pink or brick red and may have large brown patches on the wings. The larvae feed on many different trees including mangroves. This is a male.

Opodiphthera helena is found very widely from southern Queensland to Tasmania and South Australia as well as southern Western Australia. This moth is very similar to *O. eucalypti* but has a very different larva and both feed on eucalypts. *Opodiphthera helena* is the only one of the two found in southern Western Australia while *O. eucalypti* is the only one found in tropical climates. This is a male.

The enormous larva of *Coscinocera hercules* may be found in rainforest areas in northern Queensland where its favourite foodplant is the bleeding-heart tree, *Omalanthus populifolius* (Euphorbiaceae), but many other unrelated plants may be hosts. It spins a very large cocoon in the leaves of the tree.

The larva of *Syntherata janetta* is typical of the saturniids with sparse hairs, coloured turrets and a green colour. This species feeds openly on a variety of plants and the caterpillars form cocoons amongst the leaves of the foodplant.

Which is the largest moth?

Size can be measured in different ways. The moth with the largest wingspan (the distance between the tips of the forewings when the wings are spread out) is said to be a South American noctuid moth, *Thysania agrippina*. It has rather long narrow wings with a wingspan of up to 305 mm, but it is not notably large otherwise and has quite a small body.

The moth with the largest wing area is said to be the atlas moth, *Attacus atlas*, a giant saturniid or emperor moth from South-East Asia. Females of the closely related Australian and New Guinean *Coscinocera hercules* can be almost as large and one has been measured with a wing area approaching 14 000 mm^2. These moths seem like large clumsy bats when flying and make an impressive sight if many are attracted to light, as can happen during the wet season in the rainforests of northern Queensland. The atlas moth found in Darwin, *Attacus dohertyi*, is much smaller than the South-East Asian *Attacus atlas*.

The heaviest moth is the female of the Australian cossid or wood moth, *Endoxyla macleayi*, which is found from Brisbane south to near Sydney. The females are very bulky with an abdomen the size of a small banana. These moths have never been weighed fresh and the weight of a dried specimen would be puny compared to that of a live moth. They are rarely seen and most specimens have been reared from large billets from *Eucalyptus* trunks

Along with the atlas moths of Asia, the female Hercules moth, *Coscinocera hercules*, (Saturniidae), is one of the largest moths in the world—if size is measured by wing area.

or branches. The body of the female of *Endoxyla cinereus*, which has been weighed at 26 g, is a little smaller and the larva also bores in *Eucalyptus* trunks but is more widely distributed through much of Australia except Tasmania. Nothing from overseas even approaches the bulk of these moths. Their size when they emerge from the pupa is their final size; moths in their adult stage do not grow!

In contrast there are many species vying for the position of smallest moth but one of some hundreds of minute species of Nepticulidae will be the smallest. It will have a wingspan of about 3 mm and its larva will mine between the upper and lower surfaces of a leaf. We cannot suggest a species because these moths are so poorly studied that few have names.

Several other families have very small moths including the Opostegidae, Heliozelidae and Incurvariidae, but the Nepticulidae, with an almost worldwide distribution, seem to be the smallest.

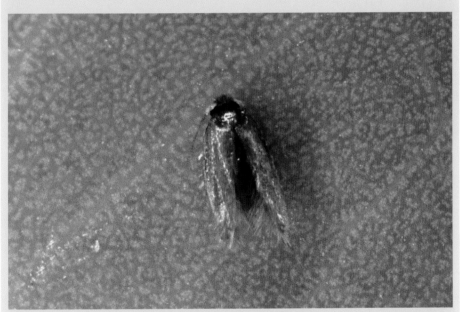

This minute nepticulid moth is one of the smallest moths in the world. Many species of Nepticulidae, from many countries, are equally tiny. This moth is about 2 mm in length.

Pollination

Not all adult moths feed at flowers and, indeed, many have reduced mouthparts or have lost them altogether. Those moths that have a fairly long, active adult life are the ones most commonly seen at flowers. The sphingids or hawk moths live a considerable time as adults, are extremely active and feed voraciously at nectar. Many noctuids also feed at nectar and some other groups of smaller moths are commonly seen at flowers too.

One difficulty in studying moth pollination is that it happens in the dark and some pollination studies of flowers have overlooked the possibility that pollination may occur at night. One moth pollination study in Australia looked at the pollination of the boab tree by hawk moths. Some species of boab trees in Africa and Madagascar are pollinated by bats and have a distinctive flower designed for bats; others are pollinated by moths. The Australian boab belongs to this group.

Flowers that are designed for moth pollination are often white or pale coloured; they are often sweetly scented and often have a fairly long corolla tube so that the long proboscis of moths is effective but the much shorter mouthparts of bees or wasps will not reach the nectaries. In flowers adapted for pollination by larger moths, the flow of nectar often peaks in the evening. Nor is it a surprise

The upper moths feeding on this daisy flower are *Glyphipterix chrysoplanetis* and the others *Glyphipterix meteora*, (Glyphipterigidae). The moths probably play a role in pollination. Photo: Peter Marriott

that moth-pollinated flowers should have a strong scent because scent is a much more effective communication device at night when airflows are stable and there are few disruptive thermals. Many rainforest flowers fit this formula.

Overseas some moth-pollinated flowers have long floral spurs but this is rare in Australia. More generalist flowers like those of eucalypts, melaleucas and daisies are visited by moths as well as many other pollinators. Australian rainforest trees with classic moth flower characteristics include *Randia*, *Canthium* and *Gardenia* in the Rubiaceae, *Hymenosporum* and *Pittosporum* in the Pittosporaceae, *Jasminum* in the Oleaceae, *Nicotiana* in the Solanaceae and *Clerodendrum* in the Verbenaceae. We have seen many hawk moths attracted to the rainforest tree *Hymenosporum flavum*, also called the native

frangipani, which has a typical pale tubular flower with a very strong sweet scent after dark.

No Australian orchids have been shown to be pollinated by moths although this is common overseas. The terrestrial orchids of Australian sclerophyll forests seem to rely mostly on wasps and flies but some rainforest or tropical woodland orchids (*Calanthe triplicata*, *Corymborkis veratrifolia* and *Habernaria* spp.) appear to have some moth-flower characteristics, but their pollinators are unknown. *Calanthe triplicata* has a floral spur 20 mm long and is almost certain to be moth pollinated.

Moths are widely thought to have flourished with the development of the flowering plants in the last 100 million years but the relationship has been more to do with larval food-plants than with pollination.

This male hawk moth, *Hyles livornicoides*, (Sphingidae), feeds at *Senna* flowers while hovering. By hovering it does not weigh the flowers down and is ready for a quick getaway as many predators frequent flowers.

The long, coiled proboscis of *Helicoverpa punctigera* (Noctuidae) extends into the floral tube of the lavender flower as the moth feeds. Like most feeding moths, it rests on the flower with wings ready for flight.

Sphingidae (Hawk Moths)

- large to very large
- smooth-scaled head and body
- wings held back near body but extended outwards and flat
- antennae simple, fairly short, often very gradually thickening until near tip
- antennae held flat, outwards from head at rest
- wings relatively narrow, hindwing much shorter than forewing
- highly streamlined moths; very agile fliers and hoverers
- very long prominent proboscis

Hawk moths are very well known and are often seen in gardens at dusk. Although they are bombycoids they have evolved in a very different direction to the other bombycoid families. They have become superb fliers and voracious feeders at nectar, hovering before a flower to feed rather than alighting. They are important pollinators, often one of the few moths that can reach the nectaries of long tubular flowers.

Most hawk moths fly at night but many fly at dusk when they can often be glimpsed speeding away against the sky. A few species in the genus *Cephonodes*, with clear wings, fly during the day.

Daphnis protrudens is a very large moth found in rainforest from the Molucca Islands to the Solomon Islands and in Queensland from Cape York Peninsula to near Townsville. The larvae have been found feeding on *Timonius timon* (Rubiaceae). This is a male.

Hawk moths lay many small eggs carefully placed on the larval foodplant and the adults can live for an extended period. The larvae feed openly on the foodplant during the day and have adopted various colour patterns like eye-spots and counter shading to avoid predation. They are easily recognised by a large, stiff, erect horn projecting from near the rear end, although a few Australian species lack this. Many species feed on the plant families Vitaceae and Rubiaceae, Balsaminaceae and Convolvulaceae although a great many other families are eaten as well. The larvae pupate in cells in the litter or soil and some have the distinctive 'jug handle' where the case of the proboscis is separate for a short distance from the rest of the pupal case.

Australia has a relatively small fauna of about 60 species, most in tropical and subtropical rainforests. There are about 1200 species worldwide.

This is a typical hawk moth larva. This species, *Theretra latreilii*, feeds on a range of plants but particularly balsam and plants of the grape family. The moth is found widely from Maluku, Indonesia and New Guinea, and south through tropical Australia, to Sydney, New South Wales.

Gnathothlibus eras is a very large moth that is widespread from the Philippines and Indonesia to the Pacific Islands. In Australia it is found in the Kimberleys, the top end of the Northern Territory and from Cape York to eastern Victoria. Its larvae feed on a range of plants but often on Vitaceae (grape family). This is a male.

The very large, common moth, *Psilogramma casuarinae*, occurs in New Guinea and in Australia in the Kimberleys, the top end of the Northern Territory, widely through Queensland, and to the south coast of New South Wales and in the Australian Capital Territory. The large, horned-at-the-rear larvae are green with cream oblique bands and feed on a very wide range of plants but in cities are usually noticed on privet and jasmine.

169

The very large *Ambulyx wildei* is found in rainforest in New Guinea and Queensland, where it occurs in the Cairns region. Nothing is known of its biology or which plants its larvae feed on. This is a male.

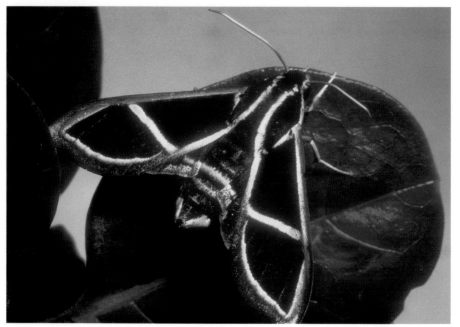

Cizara ardeniae is a large moth found from the Atherton Tableland, Queensland to Batemans Bay, New South Wales. The larvae feed on several genera of Rubiaceae, particularly on *Coprosma*. In Sydney they are common in gardens where the mirror plant, *Coprosma repens*, is growing. This is a female. Photo: Bob Jessop

THE NOCTUOID FAMILIES

The noctuoid families include all those from Oenosandridae to Noctuidae (pp. 172–197). They are characterised by thoracic hearing organs and lack of a chaetosema.

Like some of the bombycoids, some moths in this group—the subfamily Thaumetopoeinae of the Notodontidae, and the Lymantriidae—have specialised in having hairy bodies, sedentary females and a reduced or lost proboscis.

The Arctiidae often are brightly coloured and are presumed to contain poisons while the Noctuidae are strong fliers, frequently migratory, and feed avidly at flowers.

Oenosandridae

- medium-sized to large
- head and body smooth-scaled
- wings held back roof-wise over body at rest
- antennae in male sometimes simple, sometimes pectinate
- antennae held back along leading edge of forewings at rest

For many years, these moths were included in the family Notodontidae but they have been separated on characters difficult to recognise and requiring destruction of the moth to see. They have hearing organs on the thorax and share some characters with notodontids in the layout of the wing veins.

The larvae are sparsely hairy and hide during the day under bark or in crevices, emerging to feed on the foliage of trees at night. They may pupate in sheltered situations, probably in litter on the ground. Oenosandrids are night-flying and come readily to light. A few species have been reared feeding on eucalypts.

This family is only found in Australia where six species have been recorded.

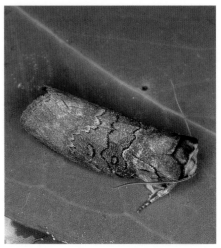

Discophlebia lucasii is a large moth that inhabits wet eucalypt forests from central New South Wales to western Victoria. Related species feed on the leaves of eucalypts.

This large male moth, *Oenosandra boisduvalii*, is playing dead; a common strategy, possibly increasing the chance of a predator losing interest. It is found from southern Queensland through southern Australia, including Tasmania to south-western Western Australia. The larvae feed on eucalypt leaves and rest during the day under bark. Photo: Peter Marriott

Notodontidae

- medium-sized to large
- those in the subfamily Thaumetopoeinae have a 'furry' or 'woolly' appearance; the remainder have rough scales or are smooth on head and body
- wings held back roof-wise over body
- antennae usually pectinate but sometimes simple
- antennae held back under forewings at rest
- hindwing shorter than forewing, with less colour pattern
- Thaumetopoeinae males have a tuft of long scales at the end of the body; females have a very large tuft of deciduous scales used to cover the eggs

Notodontids are difficult to distinguish but often have rougher scaling than other noctuoid moths. They are night-flying and come to light readily.

Moths of the subfamily Thaumetopoeinae have very hairy larvae that live in shelters of silk, or on or under bark. They pupate in litter on the ground or in the soil. The well-known processionary caterpillar, *Ochrogaster lunifer*, lives communally in a large silk bag placed either in the branches of a tree or at the base of the trunk. The larval hairs can cause severe skin rashes if the larvae or old silk bags are handled.

The larvae of the subfamily Notodontinae are not hairy but sometimes have grotesque shapes. Many species are found in tropical and subtropical rainforest, where they feed on a very wide range of plants. The genus *Hylaeora*, however, contains southern species that feed on eucalypts. Notodontines often pupate in thick, hard cocoons on the trunk of the host tree. The moth softens its cocoon with a strong alkali before emerging.

There are about 90 species in Australia and about 2800 worldwide.

Aglaosoma variegata is a large, conspicuous moth found from the Atherton Tableland, Queensland to Victoria. The large hairy larvae are found on *Acacia* (Mimosaceae) and *Banksia* (Proteaceae). Subfamily Thaumetopoeinae.

Epicoma protrahens is a small moth found in eucalypt forests from the Atherton Tableland, Queensland south to Batemans Bay, New South Wales. The larvae have been reared from *Melaleuca*. This is a male. Subfamily Thaumetopoeinae.

Neostauropus viridissimus is common in Queensland rainforest and wet eucalypt forest from Cape York to Bundaberg. It is also found in New Guinea. This large moth has a grotesque, scorpion-like larva with greatly extended true legs and an enlarged rear end, bent forward over its back with the rear prolegs forming flagella. Its foodplant has not been recorded. This is a male. Subfamily Notodontinae.

The large *Syntypistis chloropasta* is found in Queensland in rainforest and forest along streams from Cape York south to near Mackay. Nothing is known of the early stages. The leading edge of the hindwings has long scales and, when the moth rests, is placed ahead of the leading edge of the forewing. This helps to disguise the moth by hiding the hard straight line of the leading edge of the forewing. This is a male. Subfamily Notodontinae.

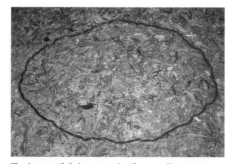

The large *Ochrogaster lunifer* is very common throughout mainland Australia. The moths are very variable in colour and may be grey, brown or yellow-brown, and often strongly striped with white. This is a male. Subfamily Thaumetopoeinae.

The larvae of *Ochrogaster lunifer*, as well as causing skin irritation, are the classical processionary caterpillars where each will follow the one in front and they may be induced to form a perpetual circle. They usually feed on wattles but sometimes on eucalypts or other trees.

This large, spectacular puss moth, *Cerura multipunctata*, is found in New Guinea and in Queensland from Lockhart River Mission south to Yeppoon. A similar species, *Cerura australis*, is found from Rockhampton to Kiama, New South Wales. The larva of *Cerura australis* feeds on *Scolopia braunii* (Flacourtiaceae) and *Polyscias* (Araliaceae). Subfamily Notodontinae.

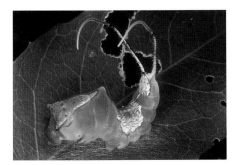

The caterpillar of *Cerura multipunctata* is found in northern Queensland rainforests. Its foodplants have not been recorded. The rear false legs are modified into filaments that the caterpillar can whip around; it can also eject formic acid from glands near its head. The puss moth of Europe is in the same genus and has a similar caterpillar. Subfamily Notodontinae.

Epicoma melanospila is a large, very common species found in eucalypt forests from Mt Molloy, Queensland through eastern New South Wales, Victoria and Tasmania to eastern South Australia. The hairy larvae have been reared on *Eucalyptus*, *Leptospermum* and *Kunzea*. The bold black spot distinguishes it from similar species. This is a male. Subfamily Thaumetopoeinae.

Hairy caterpillars and skin rashes

Long hairs on caterpillars are said to discourage bird predation although some birds are not deterred. Some caterpillars, however, have hairs of a more malignant nature that, on contact with the skin, may cause an allergic reaction in some people. Some other caterpillars always cause a reddish, raised and extremely itchy rash (also called urtication of the skin). Other caterpillars again have hairs that are short, thickened and fragile, and cause a distinct sting.

Larvae of the family Anthelidae do not usually cause urtication but their hairs may cause purely mechanical discomfort when they enter the skin in large numbers. There is even a recorded instance of blindness in one eye being caused by hairs from an anthelid larva entering the eye.

Larvae of the family Lymantriidae are more likely to be sinister. The hairs of *Euproctis edwardsii* are particularly irritating, causing severe rashes and itchiness, which may last several days. This moth is found very widely throughout Australia but not Tasmania. It is not necessary to handle the larvae to be severely affected; any disturbance nearby will cause a storm of airborne hairs to lodge on the skin. Nor is it necessary for a live larva to be involved as the old cast larval skins and the hairs associated with fresh or old cocoons can

This is the larva of *Chelepteryx collesi,* (Anthelidae), showing the stout bristles that can easily penetrate a human's skin and cause irritation. The adult moth is illustrated on p. 156. Photo: Bob Jessop

be just as effective. *Euproctis* larvae pupate under bark or in logs or timber near their host plant. If you need to move a pile of wood that may contain these larvae or cocoons, cover up with overalls, gloves, scarf and anything else, and also hose the area down thoroughly beforehand.

Many other lymantriids also cause severe urtication including *Euproctis stenomorpha* and *Euproctis lutea*, both in the Darwin area. Another lymantriid larva that can cause itchiness is *Leptocneria reducta* (which feeds on the white cedar, *Melia azedarach*), and the larvae live in a communal silken nest at the base of the tree. We thought that the lymantriid *Teia anartoides* was harmless until we reared them in the laboratory and found their final instar larvae caused considerable itchiness.

The larvae of *Panacela lewinae* (Eupterotidae), which feed on eucalypts and live in a communal silk shelter spun in the leaves of the tree, may cause severe skin rashes (as do other species of the same genus that shelter at the base of the tree). These moths are found in the damp eucalypt forests between southern New South Wales and northern Queensland.

The subfamily Thaumetopeoinae of the Notodontidae contains *Ochrogaster lunifer*, which causes severe urtication. It is widespread throughout Australia on many host plants, but most commonly on *Acacia*. Along the eastern coast it forms a communal silken shelter at the base of the tree, but inland it forms a large suspended silken bag high in the tree frequently seen on the *Acacia pendula* on the Hay Plain. The larvae leave the bag at night to feed on the leaves and after they have defoliated a tree, they set off in a long, orderly nose-to-tail line to find another. They are then called processionary caterpillars.

Ochrogaster lunifer larvae (Notodontidae) normally cluster in a large mass under a silken tent at the base of the food tree, or form a large silken bag in the branches, containing larvae, cast larval skins and droppings. Severe itching results from contact with the bag's contents or the larvae.

Some larvae of the family Limacodidae can sting. They are armed with rosettes of short, brittle spines containing an irritant, which they fold into a pouch when they are not needed but can be deployed when danger threatens. These brightly coloured slug-like caterpillars are well known; they occur on many types of trees but not all of them can sting.

Another larva that can sting is *Uraba lugens*, (Nolidae), a well-known defoliator of eucalypts found almost throughout Australia.

Lymantriidae (Tussock Moths)

- medium-sized to large
- 'furry' or 'woolly' moths with head and body well covered in long hair-like scales
- wings held back steep, or flat, roof-wise and slightly extended
- antennae broadly pectinate in the male with the pectinations extending to the tip of the antennae, which are gradually curved along their length
- antennae held back along or under the forewings
- hindwings smaller than forewings and with less colour pattern
- when they emerge from the pupa, females have a very large abdomen full of eggs ready to lay

The best way of distinguishing the Lymantriidae is the 'woolly' or 'furry' body and the short, gently curved but strongly pectinate antennae. They have a thoracic hearing organ like other noctuoids but have adopted a life strategy similar to that of many bombycoids. The females are efficient scent-producers and the males find them quickly from a long distance. The proboscis is reduced or absent.

Female tussock moths are relatively inactive; there are a few species in which the

The large *Iropoca rotundata* lives in moist eucalypt forests from central Queensland to western Victoria. Its larva has masses of pale yellow hairs and has been found feeding on eucalypts. The female has only the stumps of wings and cannot fly; it is covered in long white hairs and looks like a lump of cotton wool. This is a male.

Lymantriid larvae are hairy and many are characterised by the 'toothbrush' of hairs, here yellow-brown, standing erect about a third of the way from the head. The side-whiskers are also a feature of the larvae. Some lymantriid larvae are harmless, but others cause skin irritation when handled.

female's wings do not develop. Species with flightless females are often not fussy about the larval foodplant and may lay many of their eggs on their old cocoon.

Some lymantriid species are pests. The Gypsy moth is a very serious pest in the USA where it defoliates many species of trees and serious quarantine efforts are being made to keep it out of Australia.

The larvae of tussock moths are very hairy and can usually be recognised by the characteristic large tufts of long erect hairs sticking up from part of the body and looking like a toothbrush (hence their common name). They feed openly on the vegetation in the daytime or may hide under bark in the day and feed at night. Flimsy cocoons are spun in leaves or under bark and often the pupa can be seen through the cocoon.

Several species cause severe skin rashes in humans who come in contact with the larval hairs. Some of these are in northern Australia but the common *Euproctis edwardsii* is found throughout Australia and even loose hairs drifting on the wind can cause rashes. The hairy larvae of *Leptocneria reducta* can also cause skin irritation. They live in a silk bag at the base of their foodplant, the white cedar, *Melia azedarach* (Meliaceae), often defoliating the tree. *Leptocneria reducta* is restricted to Australia even though the tree is found from the Middle East through India to Australia. The moth has extended its range with the widespread cultivation of the white cedar and has recently become established in Canberra.

There are 70 species in Australia and about 2500 worldwide.

The genus of this species, *Artaxa epaxia*, is doubtful but the species is well known. The small moth is found in Queensland rainforest from Iron Range to the Cairns district. Nothing is known of its early stages. This is a male.

Leptocneria reducta is a medium-sized to large, very common moth found from Coen in Queensland south to Batemans Bay in New South Wales. It is found inland even to Tennant Creek, Northern Territory. The hairy larvae can cause skin irritation. They live in a silk bag at the base of their foodplant, white cedar (*Melia azedarach*). This is a male.

Lymantria antennata is a large moth found from Cooktown, Queensland to Grafton, New South Wales. The species has been reared but no details are recorded. The female has only stubs of wings and is unable to fly. This is a male.

Arctiidae (Tiger Moths)

- small to large
- smooth-scaled head and body
- antennae held out from head
- antennae simple or pectinate
- wings held back steeply roof-wise over body, flattened, or rolled about the body; the ctenuchines have their wings held out away from the body and held flat
- moths are often yellow, red, white and black

Members of this family are frequently brightly coloured and many are presumed to be poisonous or distasteful to predators. The genus *Utetheisa* contains an unknown number of species in Australia, all of which mimic each other.

Arctiids also have hearing organs on the thorax and sound-producing organs on the sides of the thorax. They produce sounds beyond human hearing but which may warn bats of their unpalatability or jam the bats' sonar. The subfamily Ctenuchinae contains mostly very similar-looking moths, probably mimicking each other. They are black and deep orange, looking rather like wasps. They are usually day-flying particularly in temperate climates but they also come to light.

Arctiid larvae are very hairy with heads that are without a colour pattern and the hairs, though usually long, are sparser than in lymantriids. They feed openly on a wide range of plants and many in the subfamily Lithosiinae (called 'footmen' in the United Kingdom) feed on lichens growing on tree trunks. Arctiids pupate among leaves, under bark or in litter or the soil; those pupating exposed often have a flimsy cocoon, easily seen through, and a few have

This large moth, *Amerila rubripes*, is found in the Kimberleys of Western Australia, the top end of the Northern Territory and in Queensland as far south as Yeppoon. When disturbed or handled, it produces a strong-smelling, acrid froth from glands near the bases of its wings that acts as a defence against predators. The larvae feed on the vine *Gymnanthera nitida* and the introduced Rubber Vine (both Periplocaceae).

a neat open cocoon reminiscent of several noughts-and-crosses grids joined up.

There are about 280 species in Australia and about 6000 species worldwide.

The medium-sized moths of the genus *Utetheisa* are distinctively coloured and also distasteful to predators. Their larvae contain poisonous alkaloids derived from their foodplants. The larvae of *U. pulchelloides* feed on boraginaceous plants, while those of the similar-looking *U. lotrix* feed on the pea genus *Crotalaria*. Both are very widely distributed in northern Australia with *U. pulchelloides* often reaching southern Australia in large numbers where it feeds in Paterson's Curse.

The medium-sized to large *Creatonotos gangis* is widespread in South-East Asia and Indonesia to Australia where it occurs in the Kimberleys, Western Australia, and the top end of the Northern Territory, and in Queensland as far south as Townsville. It is a pest species in South-East Asia where the hairy larva feeds on a very wide range of plants including grasses.

This large, brilliant moth belongs to a group of very similar species of the *Euchromia* genus. Usually day-flying, they are found in the rainforests of northern Queensland with one species extending as far south as Mackay. Their biology in Australia is unknown but several species overseas feed as larvae on Convolvulaceae.

Spilosoma canescens is a large moth that is widespread in eastern Australia from northern Queensland to Tasmania and South Australia. The larvae are very hairy and feed on a wide range of plants.

Asura monospila is a small to medium-sized moth. It is found in the rainforests near Iron Range and on the Atherton Tableland, Queensland. Nothing is known of its biology. This is a male.

There are many very similar medium sized species of the *Amata* genus in Australia, ranging over nearly the entire continent except Tasmania. They all appear to belong in a huge mimicry ring where they mimic each other and some mimic wasps. The larvae are very poorly known but probably feed on many different plants. This is a male.

The large *Nyctemera secundiana* is found from Cape York Peninsula, Queensland to Taree, New South Wales. Its larvae have been reared on the introduced weed *Crassocephalum crepidioides*, one of the daisy family. There are several very similar species of *Nyctemera*, probably forming a ring of mimics, which are all distasteful to predators. This is a female.

Asura polyspila is a small to medium-sized moth. It has been found from Mossman to near Townsville, Queensland. Nothing is known of its early stages but its larva probably feeds on dead plant detritus and lichens. This is a male.

This large, striking moth, *Oeonistis delia*, is found from Sulawesi to Samoa and New Caledonia. In Australia it occurs in rainforest from the Torres Strait Islands south to Mackay, Queensland. Its biology has not been recorded but overseas a closely related species has been found on tree trunks feeding on lichens.

Aganaidae

- medium-sized to large
- smooth-scaled head and body
- antennae usually simple, and held out from head
- wings held back, flat over body at rest
- both wings usually brightly coloured in red, yellow, orange or black
- legs are stout and fairly long

There has been much controversy over the status and classification of these moths either being placed with the Arctiidae or the Noctuidae or in a family of their own. They are brightly coloured and are presumed to be distasteful to predators. Most are active in the day but also come to light.

The larvae of *Asota* are sparsely hairy and rest openly on the leaves during the day, often two together, and look like bits of dead leaf. They feed on fig trees and pupate in a cocoon in the litter on the ground.

There are eight species in Australia and about 100 worldwide.

Digamma marmorea is a medium-sized moth that is very common and widespread from Coen, Queensland to Jervis Bay, New South Wales. It is particularly plentiful in arid areas and common at Alice Springs, Northern Territory and in the west as far south as the Pilbara. The larvae feed on the vine *Carissa ovata* (Apocynaceae). This is a male.

Agape chloropyga is a large, spectacular moth that is found from Malaysia and Borneo to Australia. Similar species are found in the Solomon Islands, New Hebrides and Fiji. In Australia they are found in Queensland south from Coen and in New South Wales as far south as Lismore. The larvae have been reared from fig trees.

The large *Asota orbona* is found in New Guinea and in Queensland from Iron Range south to Mackay. Its biology and the foodplant of the larva have not been recorded but other species of the genus feed primarily on leaves of fig trees. This is a male.

Herminiidae

- large
- smooth-scaled head and body
- wings held half back over body, flat, or extended flat on substrate
- antennae simple or pectinate
- antennae held back along leading edge of forewings at rest
- palpi often large and prominent
- wings broad, sometimes with some colour pattern on the hindwing
- drab black, grey or brown moths
- males often with one or two large scale tufts, sometimes hidden in distorted pockets on wings or on antennae or palpi

These moths are here placed in a family of their own, although in the past they have sometimes been included as part of the Noctuidae. The technical distinction from other noctuids has to do with the structure of the hearing organs and associated structures but the reality and significance of this is in doubt.

These moths have a tiny, simple, eye-like structure just above the large compound eye called an ocellus. While it is present in many moths it is particularly prominent in Herminiidae, Noctuidae and some Nolidae. It needs a good lens to see it and looks like a black ring with a tiny clear, almost crystalline centre.

Herminiids are strictly night-flying and come to light. They are moths of the forest floor and most occur in the moist forests of eastern Australia. The few larvae that are known feed on decaying plant material on the forest floor. They pupate in a cocoon in the litter.

There are 25 species in Australia but because of their checkered classification history there seems to be no estimate of the number of species worldwide.

There are several very similar species of the genus *Hydrillodes* and the females, such as this one, cannot be distinguished in photographs. These medium-sized, dingy moths are all found in eastern Australia from northern Queensland to southern New South Wales along the coast and tablelands. A very closely related species on New Caledonia has larvae that feed on plant detritus.

Nolidae

- small to large
- smooth-scaled head and body although body may have erect tufts of scales
- wings held back over body steep to flat roof-wise
- antennae usually simple
- antennae held back along the leading edge of forewings at rest
- hindwing usually without detailed colour pattern
- while some species are brightly coloured many are drab grey, small and often have numerous small tufts of scales on their wings

The classification of these moths has a long history of controversy and change. At present the family Nolidae contains moths once placed in the noctuid subfamilies Chloephorinae, Sarrothripinae and Nolinae. The Nolinae lack an ocellus but the Chloephorinae and Sarrothripinae have a typical noctuoid ocellus. All have the normal noctuoid thoracic hearing organs, and the base of the proboscis is not covered in scales.

The larvae are usually sparsely hairy and may feed openly on the vegetation or in fruit. The structure of the cocoon is one of the features of the group. It is usually spun on a leaf and tapers at both ends, and often has a ridge along the back. There are few pest species in the family with the most notable being *Earias huegeliana*, the rough bollworm of cotton, and *Uraba lugens*, the well-known skeletoniser of eucalypts.

There are 170 species in Australia and about 1400 worldwide.

Nanaguna breviuscula is common from Sri Lanka and India east to Samoa. In Australia this small moth is found in the Kimberleys, Western Australia, the top end of the Northern Territory, and from the Torres Strait Islands, Queensland to Sydney, New South Wales. Its larvae have been found on *Grevillea* (Proteaceae) and overseas on a wide range of plants. This is a female. Subfamily Sarrothripinae.

Ariola coelisigna is a small, rare species found in rainforest from Cape York to the Atherton Tableland, Queensland. It occurs also from the Himalayas to the Solomon Islands. Nothing is known of its biology. Subfamily Chloephorinae.

The medium-sized *Lasiolopha saturata* is found widely in the Oriental tropics and in Queensland from Cape York Peninsula south to near Townsville. It inhabits rainforest and monsoon forest. Nothing is known of its biology in Australia but overseas it has been reared from plants of the family Melastomataceae. Subfamily Chloephorinae.

The medium-sized *Maurilia iconica* is common from Sri Lanka to Samoa. In Australia it occurs in the Kimberleys, Western Australia, the top end of the Northern Territory and in Queensland from the Torres Strait Islands to Lamington National Park. The larvae have been recorded from a range of plants but particularly from *Terminalia* (Combretaceae). Subfamily Chloephorinae.

This small moth is a species of *Celama*. It is found in rainforest around Cairns, Queensland. Subfamily Nolinae.

The small *Acatapaustus leucospila* is found from Cairns, Queensland to Batemans Bay, New South Wales. Nothing is known of its biology. This is a female. Subfamily Nolinae.

This large moth, *Ochthophora sericina*, has a very silky sheen with otherwise drab colours. It is found from Coen to Lamington National Park, Queensland. Nothing is known of its early stages. Subfamily Sarrothripinae.

Noctuidae

- small to very large
- smooth to rough-scaled head and body
- wings may be held back over body steep roof-wise, low roof-wise or flat or with wings extended and flat against the substrate; a few adopt odd poses and one group rests with wings up like butterflies
- antennae usually simple; very rarely pectinate in male
- antennae at rest held back under wings
- wings broad, in some groups hindwing broader than forewing
- if resting with wings extended then significant colour pattern on hindwing that is not broader than forewing, if resting with wings closed then hindwing usually broader than forewing and with little colour pattern
- forewing pattern usually incorporates in some way a kidney-shaped spot about two thirds of the way along the wing in the middle and a smaller round spot at about one third of the way along the wing

This is a mega-diverse family and by far the largest family of the Lepidoptera. Consequently noctuids come in a limitless array of shapes, sizes and colours. The kidney-shaped spot and the rarity of pectinate antennae in the male are useful features in recognising them. They also have the 'noctuid ocellus': a tiny, simple, eye-like structure just above the large compound eye. It needs a good lens to see it and looks like a black ring with a tiny clear, almost crystalline centre. The base of the strong proboscis is not covered in scales.

There are two broad groups, one with three veins branching near each other on the hindwing (trifines) and one with four (quadrifines). Very broadly, most that rest with wings flat are quadrifines and those that fold their wings are trifines but there are important exceptions.

The larvae may rest openly on the vegetation in the day, they may hide under bark or on the ground and feed at night, and they may feed in flowers, buds, fruit or seeds.

The small *Eublemma cochylioides* is widespread from Africa and the Oriental region eastwards to Australia. It is found in northern Australia, south to Canberra, Australian Capital Territory, and Perth, Western Australia including the arid centre. Little is known of its biology but overseas it has been reared from a species of daisy (Asteraceae). Subfamily Acontiinae.

Alypophanes iridocosma is a small, colourful and dainty moth found in rainforest from Iron Range south to Ingham, Queensland. Nothing is known about its early stages. The resting position is unusual for a noctuid but nothing is known about why it is colourful or why it rests the way it does. Subfamily Acontiinae.

Most are not notably hairy but a few are. Some are called semi-loopers and, a bit like Geometridae, have lost some of the false legs on the abdomen and move with a looping motion. Most larvae leave the foodplant to pupate and form a silken cocoon in litter on the ground, in dead or rotten wood or in the soil.

Most noctuids are strictly night-flying but a few groups such as the subfamily Agaristinae (which includes the vine moth) are day-flying.

Many are serious pests including cutworms, army worms and earworms. The corn earworm, *Helicoverpa armigera*, and the native budworm, *Helicoverpa punctigera*, are some of the most destructive insect pests in Australia, attacking a wide range of flowers and fruit and able to migrate over long distances to attack fresh crops.

The adults of a few rainforest species pierce fruit, sucking the fruit juice and allowing mould to infect the fruit. Many others will suck fruit that is already damaged. In Asia a few of these suck tears from livestock and one will suck blood from vessels around the eye of cattle.

All noctuids are strong fliers and have a well-developed proboscis and many visit flowers. The family includes species of *Metaeomera*, the larvae of some of which feed on scale insects. It also includes the bogong moth, extraordinary for its long two-way migration and former importance as Aboriginal food (p. 203). It is one of the few noctuids with pectinate antennae in the male.

There are well over 1000 species in Australia and more than 35 000 worldwide.

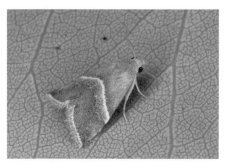

Eublemma lozostropha, a small, uncommon moth, is found from Kalbarri in Western Australia, north around northern Australia, through central Australia and down the eastern coast to southern New South Wales. Overseas some species in this genus are plant feeders but others are carnivorous on scale insects. Subfamily Acontiinae.

This small moth, a species of *Corgatha*, is found in rainforest from Cape York to Tully, Queensland. Nothing is known of its biology. It is very similar to *Corgatha flavicosta* from the Solomon Islands and further study may prove it to be the same species. Subfamily Acontiinae.

The small *Oruza stragulata* is found in New Guinea and in Australia in the top end of the Northern Territory and from Mossman to Mackay, Queensland. This is a male. Subfamily Acontiinae.

The medium-sized *Tamba cyrtogramma* occurs in rainforest from Iron Range to Babinda, Queensland. The adult moth is very variable in colour but nothing is known of its biology. Subfamily Catocalinae.

The medium-sized *Proteuxoa pissonephra* is found in south-western Western Australia. The genus *Proteuxoa* contains many similar species found across southern Australia. Only a few have been reared and they mostly fed as larvae on herbaceous plants. Subfamily Amphipyrinae.

Leucogonia ekeikei is a large, night-flying species of the usually day-flying agaristine moths like the vine moth. It is found in New Guinea and in Queensland from near Cape York south to Tully. Nothing is known of its biology. Subfamily Agaristinae.

Periopta ardescens is a large moth found in the western Kimberleys, the top end of the Northern Territory and in Queensland on Cape York Peninsula south to Coen. Its larval foodplant and behaviour are unknown. Its appearance suggests that it is closely related to the vine moth, *Phalaenoides glycinae*. Subfamily Agaristinae.

Callyna leuconota is a large moth that is found from Cooktown to Rockhampton, Queensland. Nothing is known of its biology. Subfamily Amphipyrinae.

Spodoptera picta is a medium-sized to large moth that is widespread in India, China and Japan, and eastwards to Western Samoa. It is found from the Torres Strait Islands, Queensland south to Batemans Bay, New South Wales. Its brightly striped larvae feed in lilies, including *Crinum, Clivia* and *Hippeastrum*. Other species in the genus are more sombrely coloured; their larvae are called army worms and damage lawns and greens. Subfamily Amphipyrinae.

Donuca orbigera is also a large, spectacular moth. It is common in tropical eucalypt woodlands of the Kimberleys, Western Australia and the top end of the Northern Territory, and from Cape York Peninsula south to Sydney. Nothing is known of its larvae or foodplants. This is a female. Subfamily Catocalinae.

This large, spectacular moth, *Donuca lanipes*, is found in the top end of the Northern Territory and from Cape York Peninsula south to central New South Wales with occasional stragglers as far south as Canberra, Australian Capital Territory. It is an uncommon moth throughout its range. Nothing is known of its biology. Subfamily Catocalinae. Photo: John Stockard

This is the vine moth, *Phalaenoides glycinae*. A large, day-flying moth, it is found from Brisbane, Queensland to Adelaide, South Australia, including Tasmania, and is often seen in gardens in cities and towns, feeding at flowers. Its larvae damage grape vines but also feed on native grape vines, *Cissus hypoglauca* and *Cayratia clematidea* (Vitaceae), and some other unrelated plants. This is a male. Subfamily Agaristinae. Photo: John Stockard

Speiredonia spectans is found from the islands of Torres Strait, Queensland south to central New South Wales and on Lord Howe Island. This large moth is very common and is frequently disturbed during the day in dense, shady vegetation, caves and outhouses. The foodplant of the larvae is *Acacia*. This is a male. Subfamily Catocalinae.

Rusicada revocans is a large rainforest moth widespread in the Oriental region. In Australia it occurs in Queensland south to northern New South Wales. Overseas the foodplant of the larvae is *Hibiscus* and other Malvaceae and Sterculiaceae with a New Guinea record on figs. The adult moth will suck the juices of fruit but possibly only after previous damage or decay. Subfamily Catocalinae.

The large *Eudocima jordani* is found in the rainforests of Indonesia, New Guinea and northern Queensland south to near Townsville. It is one of the important fruit-piercing moths in Queensland. This male is sucking a jackfruit. The larva probably feeds on vines of the family Menispermaceae. Subfamily Catocalinae.

Sympis rufibasis is a large moth that is widespread from Sri Lanka and India to Australia. It is found in Queensland from Cape York to near Townsville. Overseas the larvae have been found feeding on several trees of the Sapindaceae family, including litchi, rambutan and longan. This is a male. Subfamily Catocalinae.

Helicoverpa punctigera is a serious pest of many crops and is found throughout most of Australia. A medium-sized to large moth, it regularly migrates from drier areas to areas of higher rainfall, and in spring reaches Norfolk Island and New Zealand. The larva is particularly destructive as it feeds on the flowers and fruit of a huge range of plants including cotton. It is particularly fond of plants of the daisy family. Subfamily Heliothinae.

Penicillaria jocosatrix is a small moth that is common from Sri Lanka and India, east to the Pacific Islands. In Australia it is found in Western Australia south to Carnarvon, the top end of the Northern Territory and throughout the wetter parts of Queensland and New South Wales, south to Wollongong. The pose of this male is typical of the subfamily. The larva feeds on the shoots of mango trees, cashews and other Anarcardiaceae. Subfamily Euteliinae.

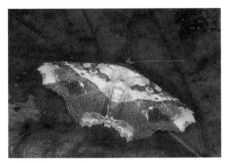

The large *Tropidtamba lepraota* is found in rainforest in New Guinea and in Queensland from Iron Range to Tully. Nothing is known of its biology. This is a male. Subfamily Catocalinae.

The large *Ophyx pseudoptera* is found in rainforest from Cape York to near Townsville, Queensland. Nothing is known of its early stages. This is a male. Subfamily Catocalinae.

Xenogenes gloriosa is a medium-sized moth of the arid areas and is common in central Australia. Nothing is known of its early stages. The bright colours of the adult suggest that the larva may feed on poisonous plants. Subfamily Catocalinae.

The large *Bastilla constricta* is found in rainforest and monsoon forest in New Guinea and in Australia from Cooktown, Queensland south to Sydney, New South Wales. Its larva has been found on *Elaeocarpus obovatus* (Elaeocarpaceae). Subfamily Catocalinae.

Metopiora sanguinata is a medium-sized moth that is much more brightly coloured than most moths in its subfamily. It is found from the top end of the Northern Territory and from the islands of Torres Strait, Queensland to Taree, New South Wales. The larvae have been found feeding on grasses. Subfamily Hadeninae.

Achaea janata is a very mobile species that is found from India to Hawaii and from Taiwan to New Zealand and Easter Island. This large moth occurs throughout Australia except for Victoria and Tasmania, but it is a migrant to most of the southern half of the continent. Its larva feeds on a very wide range of plants but is best known as the castor oil looper. Subfamily Catocalinae.

Heliothis punctifera is a very common, medium-sized moth, endemic to Australia. It is found throughout the mainland, except for Cape York Peninsula, and is an occasional migrant to Tasmania. It responds very quickly to rain in arid areas, taking advantage of a flush of ephemeral plants. Subfamily Heliothinae.

The larva of *Heliothis punctifera* feeds on many plants but particularly daisies and legumes, and can cause significant damage to crops. The larva is typical in shape and colour of many noctuid larvae that feed in flowers, buds or fruit. Subfamily Heliothinae.

Hypena conscitalis is a small moth found in rainforest and monsoon forest from Sri Lanka to Indonesia, New Guinea and in Australia from Coen south to near Townsville, Queensland. Nothing is known of its biology. This is a male. Subfamily Hypeninae.

Scriptoplusia rubiflabellata is an uncommon, medium-sized moth found in rainforest from the Bloomfield River to the Cairns region, Queensland. It is also found in New Guinea. Nothing is known of its biology but a closely related species from Sri Lanka has been reared from *Acalypha* (Euphorbiaceae). Subfamily Plusiinae.

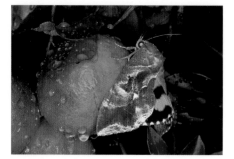

Eudocima fullonia is a large, common moth that causes much damage to fruit, particularly citrus and lychees in Queensland. The adults pierce ripe fruit. Widespread overseas and in the Pacific Islands the adults migrate. Subfamily Catocalinae. Photo: Don Sands

The larva of the fruit-piercing moth *Eudocima fullonia* feeds on menisperm vines. In the Pacific Islands, the larvae may feed on the leguminous tree *Erythrina*.

Cyclodes spectans is a large moth that is found in rainforest in New Guinea, Ambon and in Australia from Cape York south to Innisfail, Queensland. Its larva feeds on the palm *Archontophoenix alexandrae*. This is a male. Subfamily Catocalinae.

Trichoplusia orichalcea occurs from Europe to New Caledonia and New Guinea. This medium-sized moth entered Australia in the early 1970s (whether naturally or by trade is not known) and rapidly spread in eastern Australia. It is now found from the Atherton Tableland, Queensland to Victoria. In Australia its larva has been recorded feeding on soybeans but overseas from a huge range of plants and many crops. Subfamily Plusiinae.

Agrotis munda is a medium-sized to large moth that is common throughout Australia and northern Tasmania and in New Caledonia, Fiji and Tonga, and is a visitor to Norfolk Island and New Zealand. The larva is called the brown or pink cutworm and damages field crops, cereals, vegetables, lucerne and cotton. It severs the young plant at ground level and feeds on the felled plant while hiding in soft soil or litter. This is a female. Subfamily Noctuinae.

Migration

Many Australian moths are known to migrate including some important pest species whose presence in certain areas, for example Tasmania, is maintained or is greatly augmented by migration.

There are several different types of migration. The bogong moth (see p. 200) has a two-way migration where the moths migrate to mountaintops to escape unfavourable conditions, and then return to the breeding grounds when these conditions have passed. This is a very active, demanding migration in that the moths must actually fly the distance, with little or no help from the wind, and they may even have to battle against unfavourable winds. They must also maintain an approximate heading and arrive at a definite destination. Migrations like this may take weeks to accomplish but feeding is possible on the way as it is likely that they fly fairly low, within a few hundred feet of the ground. This type of migration is unusual.

Many pest species have very large one-way migrations, generally in a down-wind direction, and don't return to their breeding grounds. In these migrations, although the moths are flying, almost all of the distance covered is due to the wind. Warm westerly or north-westerly winds preceding a cold front may carry them great distances and at considerable altitude. This is a more passive type of migration and it frequently takes a shorter time, perhaps 12 to 36 hours, and no feeding is possible en route. Where the moths arrive is much more a matter of chance; some moths (including bogong moths) may frequently be caught in winds that carry them to New Zealand.

Indeed, so extensive are these movements that lepidopterists in New Zealand have set up lights on the west coast under the right weather conditions to catch moths from Australia. Losses must be enormous and there are frequent reports of vast numbers of moths washed ashore on beaches.

Spring winds carry the noctuid *Helicoverpa punctigera*, a serious pest of many crops, from central Australia to the coastal areas of south-eastern Australia, and *Persectania ewingii*, also a noctuid and pest,

This moth, *Helicoverpa armigera*, (Noctuidae), is the corn earworm or cotton bollworm, one of the most serious insect pests in Australia. The adults can migrate over great distances to find crops.

This common moth, *Persectania dyscrita*, (Noctuidae), is found in southern Australia. Its close relative, *Persectania ewingii*, migrates in spring. The larvae, called army worms, damage oat and wheat crops.

invades Tasmania each spring from areas in South Australia or western Victoria. Many other species of moths are involved in these movements. Moths migrating on weather fronts have been tracked by radar.

The biological function of these one-way flights is not clear. While it clearly permits short-term colonisation of new areas, the lack of a return flight means that any advantage from this is lost on the source population. This has led to the idea that there may sometimes be an inconspicuous return, at least from more nearby destinations, or that there are advantages in depopulating the source area where seasonally scarce larval food supplies may otherwise become depleted.

Another type of migration, frequently seen in butterflies but occurring also in some moths, is a southward movement along the eastern coast of Australia during the warmer months. This is not a mass migration and is more difficult to detect. Some of the large fruit-piercing noctuids of the tropical rainforests in northern Queensland may not be able to survive the southern winter yet some individuals of *Eudocima fullonia*, *Eudocima materna* and *Eudocima salaminia* reach Sydney in some years and even eastern Victoria. Even if these moths are several generations removed from the overwintering tropical ones it still means a substantial southward movement by each generation.

Some hawk moths have also been suspected of migrating south, the only evidence being some tropical or subtropical species that have appeared, for example, in Sydney in some years during the summer. There are no records of return flights but these would be hard to notice in moths but have also not been noticed in some conspicuous butterflies.

An intriguing possibility of migration is that some moths may reach Australia from Africa. The painted lady butterfly, *Vanessa cardui*, is a very well-known migrant normally reaching England from North Africa in numbers each year. It has several times established temporary populations on sand hills near Perth. That this has happened several times makes ship transport less likely than migration.

There is even an old record, from before the time of air travel, of the African noctuid *Grammodes stolida* appearing in Perth. Did it ride or did it fly?

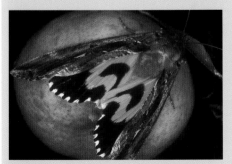

During the summer, this fruit-piercing moth, *Eudocima fullonia*, (Noctuidae), can migrate south from northern Queensland greatly extending its range down the eastern coast of Australia to Victoria.

The screen of this centimetric radar set shows a band of images of migrating moths. Photo: John Green

The Bogong Moth

The bogong moth, *Agrotis infusa*, is one of the best-known moths in south-eastern Australia and excites human interest for several quite different reasons. The genus *Agrotis* is found worldwide and other species overseas are known to be migrants. Unusually for members of the family Noctuidae, *Agrotis* males have pectinate antennae—a useful identification point when moths get themselves baked into loaves of bread or drowned in take-away food as sometimes happens. The larvae are known as black cutworms and cause minor damage to vegetable and cash crop seedlings. The cutworm larva lives in the surface layer of the soil and at night fells a young plant by chewing it off at ground level and it then proceeds to consume the felled plant from the safety of the soil.

The bogong moth is found throughout southern Australia south of the Tropic of Capricorn, with large populations on the western slopes and plains of southern Queensland, New South Wales and Victoria. In August and September, adult moths migrate south or east, sometimes over thousands of kilometres to mountaintops in the Snowy Mountains of New South Wales and the Victorian Alps. These migrations involve huge numbers of moths,

The bogong moth, *Agrotis infusa*, was a good source of food for Aboriginal Australians because of its high fat content and because the moth could be easily obtained in very large numbers. The black streak on the wing between the two spots is the best way to distinguish it from other species of *Agrotis*.

and they can cause problems in the cities through which they travel where they are attracted to lights and enter homes. In about October, they reach the mountains and then move slowly up to the highest peaks where they will rest over summer in caves, cracks and crevices in the rocky summits. Here it is cool and moist, and the moths try to select the crevices most sheltered from the wind; in the caves they pack in very closely together like tiles, on the rock faces. Here they are prey to many animals, particularly the Australian raven, foxes, and also the mountain pygmy possum. While in the caves most seem to remain resting but some will fly at night and the dusk flight around a mountaintop can be spectacular.

They remain in the mountains until late February when they commence their journey back to the breeding grounds. Both males and females migrate but the females' ovaries do not develop until the return migration so that they cannot lay eggs until they return to the plains. When the moths leave the mountains in autumn, flowers are scarce, but the return migration and egg maturation have to be fuelled. Ken Green of the New South Wales National Parks and Wildlife Service has observed bogong moths feeding avidly on the honeydew from lerp insects, and believes this sugar source may play an important role in the moths' biology.

On their return to the breeding grounds the females lay their eggs, which hatch after a

While aestivating in caves, bogong moths press together in a tile-like pattern to conserve moisture. Vast numbers can rest in narrow crevices when packed in so tightly. Photo: Ian Common

week or two and the larvae develop over the winter. The larvae pupate in the soil in late winter emerging after a few weeks, ready for the next migration in August or September. However, in the wetter areas of the coast and tablelands of southern Queensland, New South Wales and Victoria, not all moths migrate. Some will remain and breed in the spring producing other generations that eventually lay eggs in autumn for the next winter generation.

It is not known if the moth populations that migrate are separate from those that remain or whether a single population can do either depending on climatic cues. The non-migratory generation of moths has paler hind-wings and is easily distinguishable from all moths flying in the spring. Migration is thought to be a strategy to allow the survival of the adult moths in a cool moist place when the breeding grounds become too hot and dry for their survival over summer. Most other moths solve this problem by having a pupal resting stage over summer in some sheltered place or perhaps beneath the soil. The bogong moths in Western Australia are not known to migrate and their summer survival strategy is unknown.

Bogong moths are fairly small with a wingspan of about 45 mm, and to survive their summer visit to the mountains they lay down large reserves of fat in the abdomen. Because they occur in such vast numbers at the mountaintops they are very easy to collect. Each year

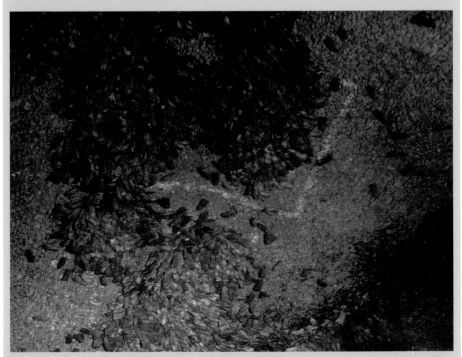

The walls of even the most exposed caves are partly covered in moths. In years when bogong moths are less plentiful the moths are concentrated in the deepest crevices and are not so easily visible. Photo: Rob Carrick

in early summer the surrounding Aboriginal tribes congregated to harvest the moths. The moths were gathered by disturbing the clustered masses allowing them to fall by their hundreds into a bag held below. Alternatively a stick could be run up a crevice and the moths would pour out the base of the crevice like water. The moths were usually cooked on the spot; after the wings and legs were partly singed off, they were eaten directly or pounded into cakes. So rich is this fatty diet that Aborigines were often sick until they adjusted to it. The moths were such an important seasonal food source that the tribes migrated considerable distances to collect them, and customs involving tribal contacts were reserved for this season.

There are records from the nineteenth century of hundreds of moths swept from churches or washed up on beaches but the main problem with moths invading buildings began with the ready availability of cheap electric lighting. Every spring there are reports of hundreds of moths attracted to well-lit buildings where they soon die of desiccation in the dry air inside the building. These moths are not clothes moths and their larvae will not eat woollens or dried foods, but their scales can soil curtains and floor coverings, and their dead bodies are a source of food for other undesirables such as cockroaches, carpet beetles and mice. To minimise the entry of moths into buildings, external lights should be yellow wherever possible (sodium vapour lights or specially made, yellow fluorescent tubes are fine). External and internal lights should not be left on all night as this allows moths no time to escape. If a light needs to be on at night it should be switched off in the early hours, if possible, to allow the moths to disperse before dawn when they seek the dark interior of the building to hide during the day. To prevent entry to a building, wire mesh with about 4 mm gaps should be fitted to air-conditioning intakes and other gaps should be meshed or sealed as far as possible.

In 1966 the mountain pygmy possum, previously known only from fossils, was found alive in the Victorian Alps and later in the Snowy Mountains. Researchers discovered that female possums in particular are often found in boulder fields where bogong moths are plentiful. The possums eat many things but a correlation has been suggested between lack of breeding success and poor bogong moth years. It seems possible that the possums, already faced with problems of climatic warming, may also be vulnerable to any possible decline in the moth supply through human activities, either in the breeding grounds or through the expansion of well-lit towns and cities on the migration path.

In 2001 Ken Green observed grass dying around the bogong moth sites on mountaintops on the Snowy Mountains. Investigation revealed high arsenic levels in the soil and in the moths. More research is needed to find out where this arsenic is coming from and what the implications are for the moths and for the predators, possums or people, eating the moths. It may also be an interesting case of a pollutant being carried into the heart of a wilderness area by natural processes.

A frequent question is, 'What do bogong moths taste like?'. It is reported that they have a 'nutty taste'. Lepidopterists face a few hazards but one of the worst is to develop an allergy to the moths we study. To minimise unnecessary exposure we have not eaten bogong moths (though few people believe this excuse). This answer is clearly always a disappointment.

GLOSSARY

Aestivation   A summer resting period in the life cycle of an insect. It is a summer equivalent of hibernation.

Antenna (plural, antennae)   The long, paired, sensory, thread-like or feathery organ which arises near the top of the head.

Biological control   A procedure where an organism from somewhere else is introduced and released in the hope that it will help control a pest or weed.

Borer   A larva which excavates a large tunnel in wood, bark or fruit.

Cell   The central area of the wing, which is surrounded by veins.

Chaetosema (plural, chaetosemata) A pincushion-like organ with radially projecting hair-scales found just above the eye and behind the base of the antenna. It is very difficult to see without a microscope but very useful in distinguishing moth families.

Cocoon   A silken enclosing shelter made by a larva when preparing to turn into a pupa.

Complete metamorphosis   An insect life cycle, which has the following stages: egg, larva, pupa and adult. The stages look very different from each other. This contrasts with a life cycle that has an egg, a series of nymphs and an adult, called an incomplete metamorphosis. Here, the nymphal stages look increasingly like the adult.

Foodplant   The plant upon which the larva feeds. In this book the plant that the adult feeds on is called a nectar source to avoid confusion. The term 'host plant' is sometimes used but is less precise and could refer to another type of association.

Forewing   The front wing of the pair on each side of a moth.

Frass   The droppings of a larva.

Frenulum   A bristle (males) or several bristles (females) arising near the base of the hindwing and meshing with a hook or patch of scales on the underside of the forewing (see p. 35).

Hindwing   The back wing of the pair on each side of a moth.

Instar   The stage between moults in the growth of a larva. The first instar is the stage between the egg and the first moult. Many larvae have five instars; they shed their skin four times to become bigger larvae and once to become pupae.

Lanceolate   Shaped like an elongate oval, three to six times as long as broad, tapering to a tip. A common shape of the hindwing.

Larva (plural, larvae)   The grub or caterpillar stage of the life cycle.

Lepidoptera   The insect order that includes moths and butterflies but not other insects.

Macrolepidoptera   Moths with a wingspan generally over 3 cm. The term is applied to moths of all the families after the Pyralidae in the book and includes the families before the Pyralidae listed under microlepidoptera. Pyralidae are anomalous and are treated as microlepidoptera by some authors and as a separate group 'pyralids' by others.

Metamorphosis   The changes in form occurring during the progression of a moth through its life cycle from egg to adult. Sometimes used as another term for life cycle.

Microlepidoptera   Small moths with a wingspan usually less than 3 cm. The term is applied to all moths in particular families where most species are small. All families before the Pyralidae in the book except

Hepialidae, Cossidae, Dugeoneidae, Castniidae, Hyblaeidae and Thyrididae are included in the microlepidoptera. It is a useful informal group and not part of the classification.

Mimic   In a restricted sense, any species that resembles or imitates another unrelated species. A mimic may resemble a model because the model is distasteful, poisonous or predatory and the mimic derives protection from its resemblance. When mimic is used in this sense then species which resemble an unrelated species show mimicry and those which imitate objects are said to show 'protective resemblance' or 'crypsis'. In a broad sense, a species which resembles or imitates another species or an object.

Mimicry ring   A group of species, some of which are not related, that resemble each other. It implies that the species derive benefit from their similarity.

Mine   A tunnel or burrow a larva excavates when feeding in plant tissues. It may be a blotch, a blister or serpentine and may be in a leaf, petiole, bark or wood. It is made by a small larva; the burrow made by a large larva in wood is called a bore.

Miner   A larva which makes a mine.

Ocellus (plural, ocelli)   A very small domed structure found just above the eye in adult moths, usually with a single tiny lens. It is a light-sensitive organ difficult to see without a microscope.

Palps   The scaled, protruding appendages in front of the head. In this book the labial palps are called palps, as they are the most visible. There are shorter maxillary palps but they are often absent and are less visible when present.

Pectinate antenna   An antenna that looks 'feathery' or 'comb-like' because there are numerous short branches arising along all or part of its length.

Pheromone   A chemical produced by a moth to communicate with another individual of the same species. Usually produced by the female to attract males and disseminated

in the air, but sometimes produced by the male when close to the female to identify the male as the correct species.

Proboscis   The usually long, coiled tube forming the mouthparts of the adult moth through which it sucks liquids.

Prolegs   The fleshy legs arising from some of the abdominal segments of a larva. These are distinct from the three pairs of true legs arising from the thorax.

Pupa (plural, pupae)   The stage of the life cycle between larva and adult, during which the tissues are completely reorganised in the metamorphosis from the larva to the adult.

Pupate   When the larva sheds its skin to become a pupa.

Retinaculum   A hook-like flap or a patch of scales on the underside of the forewing which meshes with the frenulum (see p. 35).

Scales   The flattened, modified hairs that usually cover the wings and body of a moth.

Silk   The strong flexible thread produced by modified salivary glands of a larva. Silk is used for several different purposes but most notably to construct the cocoon.

Sinuate   Wavy, applying particularly to the margin of the wing but sometimes to a leaf mine. When applied to Gelechiidae it means that the outer margin of the hindwing has a single wave in it, i.e. the apex is pointed and the margin goes from convex to concave to convex. The concavity may be marked or barely discernable.

Spiracle   The opening in the body wall through which gasses pass in respiration. A small round or oval organ on each side of almost every segment of the moth found in all life stages except the egg (see picture, p. 7).

Vein   One of the tubes that strengthen the wings of moths. They are also used to carry fluid when the wings are expanded after emergence from the pupa.

INDEX